Springer Series on
Atoms+Plasmas

20

Guest Editor: Prof. J. P. Toennies

Springer

Berlin
Heidelberg
New York
Barcelona
Budapest
Hong Kong
London
Milan
Paris
Santa Clara
Singapore
Tokyo

Springer Series on
Atoms+Plasmas

Editors: G. Ecker   P. Lambropoulos   I. I. Sobel'man   H. Walther
Managing Editor: H. K. V. Lotsch

N. Stolterfoht   R. D. DuBois
R. D. Rivarola

Electron Emission in Heavy Ion–Atom Collisions

With 78 Figures

 Springer

Priv.-Doz. Dr. Nikolaus Stolterfoht

Hahn-Meitner-Institut Berlin,
Glienickerstraße 100
D-14109 Berlin, Germany

Professor Robert D. DuBois

University of Missouri-Rolla,
Department of Physics
Rolla, MO 65401, USA

Professor Roberto D. Rivarola

Universidad Nacional de Rosario,
and Instituto de Fisica Rosario, CONICET,
Rosario, 2000, Argentina

Guest Editor: Professor Dr. J. Peter Toennies

Max-Planck-Institut für Strömungsforschung, Bunsenstrasse 10,
D-37073 Göttingen, Germany

Series Editors:

Professor Dr. Günter Ecker

Ruhr-Universität Bochum, Lehrstuhl Theoretische Physik I, Universitätsstrasse 150,
D-44801 Bochum, Germany

Professor Peter Lambropoulos, Ph. D.

Max-Planck-Institut für Quantenoptik, D-85748 Garching, Germany, and
Foundation for Research and Technology – Hellas (FO.R.T.H.),
Institute of Electronic Structure & Laser (IESL) and
University of Crete, PO Box 1527, Heraklion, Crete 71110, Greece

Professor Igor I. Sobel'man

Lebedev Physics Institute, Russian Academy of Sciences,
Leninsky Prospekt 53, 117924 Moscow, Russia

Professor Dr. Herbert Walther

Sektion Physik der Universität München, Am Coulombwall 1,
D-85748 Garching/München, Germany

Managing Editor: Dr.-Ing. Helmut K.V. Lotsch

Springer-Verlag, Tiergartenstrasse 17, D-69121 Heidelberg, Germany

Library of Congress Cataloging-in-Publication Data applied for.
Die Deutsche Bibliothek – CIP-Einheitsaufnahme
Stolterfoht, Nikolaus: Electron emission in heavy ion–atom collision / N. Stolterfoht; R. D. DuBois; R. D. Rivarola. – Berlin;
Heidelberg; New York; Barcelona; Budapest; Hong Kong; London; Milan; Paris; Santa Clara; Singapore; Tokyo: Springer, 1997
(Springer series on atoms + plasmas; 20) ISBN 3-540-63184-4

ISSN 0177-6495
ISBN 3-540-63184-4 Springer-Verlag Berlin Heidelberg New York

Cover design: *design & production* GmbH, Heidelberg
Typesetting: Data conversion by Lewis T$_E$X, Berlin

SPIN 10632964   54/3144 - 5 4 3 2 1 0 - Printed on acid-free paper

To
Gene Rudd

Preface

This book is devoted to 30 years of lively research that has taken place in the field of ionization in ion–atom collisions. The ionization process, where an electron is liberated to the continuum of an atom, is of vital interest for basic research and many applications such as plasma physics, astrophysics, and radiation physics. The book is a broad review of theoretical and experimental studies addressing electron emission in energetic ion–atom collisions. High incident energies are considered, where the projectile velocity is greater than the initial velocity of the electron participating in the emission process. At high energies the projectile interactions may be considered as a relatively weak perturbation. However, with the advent of highly charged projectile ions, strong interactions can occur in energetic ion–atom collisions. The book puts emphasis on the interpretation of different ionization mechanisms to support a detailed understanding of processes leading to electron emission. Hence, it is written for researchers interested in applications and students learning fundamental aspects of ion–atom collisions. It addresses both experimentalists searching for information about the experimental apparatus and theorists who want to enter into the field of collision physics. It is assumed that readers interested in the theoretical methods are familiar with the basics of quantum mechanics.

The review treats collisions involving incident bare ions as well as projectiles carrying electrons. The ionization mechanisms are classified in terms of the strength of the electronic interaction with heavy particles. These mechanisms are interpreted in terms of Coulomb centers formed by the positive charges of the projectile and target nuclei, which interacts strongly with the outgoing electron. In particular, the cases of electron emission in one and two centers are analyzed. As examples, electron emission in binary and soft collisions with bare projectiles are treated as cases for single-projectile and target centers, respectively. General properties of two-center electron emission are analyzed by using electron capture to the continuum and saddle-point electron emission as specific examples. For dressed projectiles, particular attention is devoted to screening effects, anomalies in the binary-encounter peak production, diffraction effects, and dielectronic processes involving two active electrons. Lastly, a brief overview of multiple ionization processes is presented. The survey concludes with a compilation of experimental studies related to heavy ion impact ionization. Publications that appeared before the end of 1996 are taken into consideration.

This work has been prepared in close communication with many colleagues who are too numerous to acknowledge individually. Their assistance and comments contributed greatly to this work. Our deepest thanks go to Eugene Rudd whose research formed the foundation for this field of study and whose ideas guided us at various stages of this review. We are much indebted to Wolfgang Meckbach, Antoine Salin, Theo Zouros, Bela Sulik, Mitio Inokuti, Victor Ponce, Gregor Schiwietz, John Tanis, and Carlos Garibotti who critically read parts of the manuscript and provided very valuable comments. We thank Laszlo Gulyás for his assistance in using the CDW-EIS code. Numerous discussions and communications with Carlos Reinhold, Joe Macek, Dave Schulz, Joachim Burgdörfer, Yasu Yamazaki, Pablo Fainstein, Knud Taulbjerg, Jorge Miraglia, and Ron Olson are gratefully acknowledged. R. D. Rivarola acknowledges Silvia Corchs for her valuable assistance in preparing part of the manuscript. R. D. Dubois is indebted to the Hahn-Meitner Institute for support during his stay in Berlin. N. Stolterfoht is greateful for support by the Instituto de Fisica during his visit to the Universidad Nacional de Rosario.

Berlin, Rolla, Rosario
August 1997

N. Stolterfoht
R. D. DuBois
R. D. Rivarola

Contents

1. Introduction

For several decades the fundamental process of electron emission in ion-atom collisions has received considerable attention. Ionization plays vital roles in various fields of applied physics such as plasma physics, astrophysics, and radiation physics. Hence, understanding the basic atomic ionization mechanisms has practical, as well as fundamental, importance. Fascination with break-up events in ion-atom collisions is generated largely by the intellectual challenge of studying the dynamics of many-body problems. The motion of an electron, initially bound to a target atom, liberated during a collision with a projectile ion and finally interacting with both collision partners is a highly complex phenomenon. Since the nature of the Coulomb interaction is known, the understanding of the mechanisms relevant for ion-induced electron emission promises important insight into the many-body problem. Electron emission is very effective in providing detailed information about the structure of atoms and the dynamics of collision mechanisms.

The description of ionizing collisions is basically a three-body problem whose exploration represents a great challenge for theorists since the three-body problem underlying the process of ion-induced electron emission can only be solved by approximation. A well-known method to treat ionization in ion-atom collisions is to reduce the three-body problem to a corresponding two-body problem or to a sequence of two-body interactions. Furthermore, the three-body problem is frequently solved using first-order perturbation theory. Attempts to treat the process of electron emission by higher-order theories are relatively rare and generally require considerably more effort than conventional methods.

Our basic understanding of electron production in ion-atom collisions originates from the fundamental work of Rutherford (1911) who succeeded in describing collisions between charged particles at the beginning of this century. Since it represents an analytic solution for the two-body problem where the incident projectile interacts with a target electron initially at rest, Rutherford's formalism is of interest to and has permeated many branches of collision physics. Classical models, based on Rutherford's theory, include describing the two-body process of a projectile colliding with a moving electron, i.e., the binary-encounter theory, [*Gryzinski* (1959, 1965)]. Furthermore, the Rutherford formula enters into the classical trajectory Monte Carlo method

(CTMC) which includes interactions of the electron with both the projectile and the target nucleus [*Abrines* and *Pecival* (1966a,b); *Olson* and *Salop* (1977)]. Therefore, the CTMC is suitable for describing the three-body interactions leading to ionization.

Since its beginning, quantum mechanics has been used extensively in the description of ion-atom collisions. In the early 1930s Bethe performed pioneering work to describe electron emission by charged particles by means of the (first) Born approximation. The Born approximation is based on perturbation theory, which is applicable for weak projectile interactions. However, the interaction with the target nucleus is fully taken into account. In this work, ionization mechanisms are classified by using the concept of centers, where a center is formed by a heavy particle that interacts strongly with the outgoing electron. Therefore, the Born approximation describes essentially the emission of the electron from a single atomic center formed by the target nucleus.

In spite of the considerable interest in ionization processes, only a few measurements of ion-ejected electrons existed before the 1960s because of difficulties with the detection of low-energy electrons. Adequate measurements of ejected low-energy electrons, which predominate in ion-atom collisions, require great effort by experimentalists. *Blauth* (1957), and *Moe* and *Petch* (1958) were the first to measure electron production in ion-atom collisions. In the 1960s *Kuyatt* and *Jorgensen* (1963) and *Rudd* and collaborators (1963, 1966a,b) started systematic measurements of nearly complete energy and angular distributions of ejected electrons. These pioneering experiments were unique for nearly a decade and revealed many of the processes that have been extensively studied since that time. For example, electron production mechanisms in soft collisions (SC) and binary encounters (BE), corresponding to large and small impact parameters, respectively, were studied in detail. Furthermore, the process of electron capture to the projectile continuum (ECC), producing a cusp shaped peak at $0°$, was observed [*Crooks* and *Rudd* (1970)]. The ECC mechanism provides an outstanding example of two-center electron emission, since electron capture is governed by the long-range Coulomb forces of both collision partners.

In the early 1970s, experiments of electron emission were carried out at other laboratories by *Stolterfoht* (1971) and *Toburen* (1971) using projectiles with higher incident energies. *Burch* et al. (1972, 1973) used heavy ions from a tandem Van de Graaff accelerator which provided projectile energies of the order of 50 MeV. These heavy projectiles are generally "dressed" as they carry electrons of their own into the collision and give rise to characteristic collision effects. Interactions involving dressed projectiles were theoretically studied by *Bates* and *Griffing* (1953) in the early 1950s. They demonstrated that one of the ionization channels involves a projectile electron remaining in its ground state during the collision where its main effect is to screen the projectile nuclear charge. There is also a second ionization channel where the projectile electron is removed from its ground state. This channel may contribute independently of the projectile to the ionization of the target atom.

A collision involving a dressed projectile can also result in the removal of a projectile electron due to its interaction with the target nucleus. This case refers to a reversed collision system in which the projectile takes the role of the target and vice versa. Electrons ejected from the projectile may produce significant contributions in measured spectra. Due to kinematic effects, the projectile electrons are concentrated in a single maximum, i.e., the electron-loss to the continuum (ELC) peak, observed in the early 1970s [*Burch* et al. (1973), *Wilson* and *Toburen* (1973)]. At high incident energies the ELC peak is well resolved from the contribution of the target electrons and can therefore be studied separately.

As more and more experimental data were measured, it became clear that one-center models, such as the Born approximation, are not sufficient in describing ionization in ion-atom collisions. In particular, a breakdown of the one-center picture is expected as the outgoing electron is influenced by the Coulomb fields of both the target and projectile. Indications for two-center effects had already been found in earlier investigations by *Rudd* et al. (1966a) where an enhancement of the forward electron emission was observed. This finding led to the discovery of the cusp-shaped ECC peak mentioned above.

In the 1980s increasing attention was devoted to studies of two-center effects in ionizing collisions. Groups at Bariloche (Argentina), Debrecen (Hungary), and Oak Ridge (USA) studied the details of the ECC cusp at $0°$ to verify theories treating ion-atom collisions in higher orders. Further experimental work by *Stolterfoht* et al. (1987) and *Pedersen* et al. (1990), using fast highly-charged projectiles, have shown that two-center effects are not restricted to forward angles but are important in the whole angular range of the ejected electrons. These effects were confirmed theoretically by *Fainstein* and collaborators (1987, 1988a) using a continuum distorted wave method. Furthermore, controversial results were obtained by several groups concerning electron emission from the saddle point, formed by the two-center field of the collision partners [*Olson* (1983), *Gay* et al. (1990), *Meckbach* et al. (1991), *DuBois* (1993)]. The work about two-center effects created lively discussions that have continued until the present time.

The present work reviews research concerning electron production mechanisms in ion-atom collisions performed during the past 30 years. Because of the extensive amount of research in the field, the present review of electron production mechanisms must be limited. Earlier work which has been described by *Rudd* and *Macek* (1972), *Rudd* (1975), *Stolterfoht* (1978), and *Berényi* (1981), is not discussed in detail. Studies involving proton impact are included only in specific cases. They are not fully treated here since these have recently been reviewed by *Rudd* et al. (1992). Furthermore, the emission of Auger electrons is not discussed, as this subject lies outside the scope of the present work which concentrates on continuum electrons. Summaries dealing with the atomic structure aspect are given by *Mehlhorn* (1985) and *Stolter-*

foht (1987a). Recent reviews related to the present work are provided by *McDaniel* et al. (1993), *Rudd* et al. (1996), *Lucas* et al. (1997) and *McGuire* (1997).

Apart from its limitations, this work provides a broad overview of the mechanisms of electron emission in energetic ion-atom collisions. In this sense, the present work goes beyond most previous reviews which were limited to specific aspects of electron emission in atomic collisions. In particular, both theory and experiment are presented with an attempt to provide a rather complete summary of the current literature. In the discussion of ionization mechanisms, attention is focused on heavy ions which produce strong perturbations of the target atom. Nevertheless, lighter projectiles such as helium ions and atomic hydrogen are also important subjects in the present discussion. Numerous studies, which have been performed using helium ions and atomic hydrogen as a dressed projectile, are of particular interest in this review. Emphasis is given to fast collisions, for which the velocity of the incident projectile is larger than the velocity of the bound electron to be removed. Often, helium is considered as the target.

The present work is structured in terms of increasing complexity of collision systems with varying numbers of particles and interactions. This structure is incorporated into the discussions of both theory and experiment. First, the relatively simple case of two-body interactions is presented. Then, the three-body problem is discussed, first by treating the projectile-electron interaction in first-order, followed by treatments of this interaction in higher order. Finally, we consider the four-body problem involving dielectronic processes where the interaction between the electrons plays an important role. The discussion of the electron production mechanisms is accompanied by intuitive pictures of viewing electron emission from collision centers formed by the strongly interacting heavy particles. We already introduced the cases of electron emission from a single target center and from the two-centers formed by the projectile and target nuclei. Additional information about the center concept is given in the following section.

2. Overview of Ionization Mechanisms

To introduce the basic feature of the ionization mechanisms, in Fig. 2.1 we show a relatively simple electron spectrum indicating various characteristic phenomena. The data refer to electron emission cross sections calculated for the collision system 25-MeV/u Mo^{40+} + He at an observation angle of 5° with respect to the incident beam direction [*Stolterfoht* et al. (1987)]. The theoretical method used in the calculations and the spectral structures of the results will be discussed in detail in the next sections, therefore, only a short description will be given here. The low-energy region is attributed to electrons produced in soft collisions (SC). At the high-energy end of the spectrum the pronounced peak is due to binary-encounter (BE) collisions. The structure labeled ECC is associated with the process of electron capture to the continuum. The region between the ECC peak and the SC shoulder is attributed to two-center electron emission (TCEE). This region is rather extensive, particularly for high-energy projectiles.

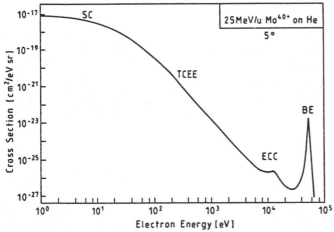

Fig. 2.1. Electron spectrum demonstrating the different mechanisms for electron production: Soft collision (SC), two-center electron emission (TCEE), electron capture to the continuum (ECC), and binary encounter (BE). The data referring to electron emission at 5° in collisions of 25-MeV/u Mo^{40+} on He are calculated by means of the CDW-EIS approximation [*Fainstein* et al. (1988)]. From *Stolterfoht* et al. (1987)

2.1 Considerations to the Center Picture

As various mechanisms can produce electrons in heavy ion-atom collisions, it is useful to find general categories under which they can be summarized. In this work, the electron production mechanisms will be discussed in terms of Coulombic centers which are associated with the heavy nuclei of the collision partners. The formation of a center involves a strong interaction of the nucleus with the active electron so that a description beyond first-order perturbation theory is required. The center picture takes into account the specific feature of an ion-atom collision system that consists of two heavy particles and one, or more, electrons. Instead of regarding the complex dynamics of a many-body system with particles of equal masses, one concentrates on the electronic interaction with heavy centers whose motion is practically independent of the light electrons. In recent years the center picture has been recognized as a useful concept, e.g. see the various contributions by *Schultz* et al. (1996), *Rivarola* et al. (1996), *Stolterfoht* (1996), and *Richard* (1996) to the symposium for two-center effects in ion-atom collisions held in 1995 at Lincoln, Nebraska.

Formally, centers can be defined in terms of the final-state interaction of the collision partners with the active electron. As noted already, the first Born approximation includes the full target interaction in the final state and is thus associated with the target-center case. Likewise, the inclusion of higher-order projectile-electron interactions in the final state corresponds to the projectile-center case. Moreover, theories including two-center wave functions in the final states refer to the two-center case. The comprehension of the center concept in terms of final state interactions is convenient for quantum-mechanical treatments. However, other descriptions are relevant for classical theories. For instance, the binary-encounter theory is associated with the projectile-center case and the classical-trajectory Monte-Carlo method refers to the two-center case.

A few comments shall be added to elucidate the center picture. A nucleus becomes a center only during the collision, i.e., we disregard the pre-collisional interaction of the active electron with the nucleus at which the electron is bound before the collision. This is done to allow for the consideration of the zero and single-center cases where the collision partner, at which the electron is initially bound, acts as a small perturbation. For instance, when target ionization is treated, one may consider the projectile-center case where the target nucleus plays a minor role.

Neglecting the collisional effect of the nucleus, to which the electron is initially bound, is known as the free-electron or impulse approximation. For example, the impulsive approach is adopted in the binary-encounter theory. Furthermore, this approach is implemented in a simplified version of the Born approximation in which a plane wave is used for the electronic final state. This plane wave describes an outgoing electron whose interaction with the nuclei is neglected. Hence, the free-electron Born approximation contains neither a projectile nor a target center and, hence, it will also be referred to as zero-center case. This free-electron Born approximation differs from the binary-encounter theory which contains a projectile center. Otherwise, the theoretical models are closely related [*Madison* and *Merzbacher* (1975)].

In two-body collisions the difference between the Born approximation and the binary-encounter theory is relevant only when dressed projectiles are treated. For bare projectiles a violent interaction may theoretically be treated as a weak perturbation. This is due to the exceptional property of the Coulomb potential which permits calculating accurate scattering cross sections for binary collisions by first-order perturbation theory. This unusual property will be commented on in more detail later. Here, in view of the center concept, it should be kept in mind that the Born approximation (which treats the projectile interaction in first-order) fully describes a center formed by a bare projectile. In contrast, for dressed projectiles the Born approximation is not sufficient, i. e. it is necessary to apply a theory involving a projectile-center, such as the binary-encounter theory. Also, it is useful to separate the treatment of bare and dressed projectiles. This will be done throughout the present review.

A few aspects of the center concept are schematically displayed in Fig. 2.2 showing the removal of an electron in a perturbative interaction with the incident ion. We consider the cases where the outgoing electron is unaffected, deflected in the field of the target center, and deflected by both target and projectile centers. These cases correspond to a binary collision without center, electron emission involving a single-target center and electron emission involving two centers, respectively. Furthermore, Fig. 2.3 contains a scenario of electron emission events that will be taken as a guidance for the discussion in the following sections.

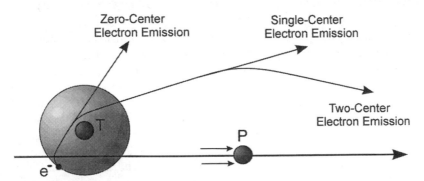

Fig. 2.2. Mechanisms for electron emission associated with different trajectories. The zero-center case refers to a weak two-body interaction between the projectile and the target electron. In single-center electron emission the electron is affected by the nuclear field of the target. In two-center electron emission the electron is affected by the field of both the target (T) and projectile (P) nuclei

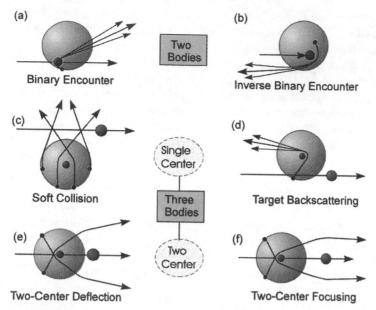

Fig. 2.3a–f. Scenario of electron production mechanisms for which different centers are important. As described in the text, a center implies a strong interaction between the associated nucleus and the active electron so that higher-order terms become important. In (**a**) a fast projectile hits a slow target electron in a binary collision where the projectile may form a center depending on its interaction strength. In (**b**) a fast electron leaves the target as it is scattered by the quasi static field of a slow projectile. In (**c**) and (**d**), respectively, a slow and a fast electron is scattered by the target nucleus which forms a center. In (**e**) and (**f**) both the target and projectile nuclei form centers where the outgoing electrons are significantly scattered

2.2 Two Bodies: Binary-Encounter Electron Emission

A relatively simple interaction mechanism leading to ionization is due to a projectile interacting with an electron in a binary collision (Fig. 2.3a). This type of interaction corresponds to a two-body process. If the interaction of the outgoing electron with the target nucleus is neglected, the ionization process is referred to as binary-encounter electron emission, which may be described by a classical theory [*Gryzinski* (1959), *Bonsen* and *Vriens* (1970)]. As noted already, binary-collisions are generally associated with the projectile-center case. The target nucleus is responsible for the initial velocity distribution of the electrons, but otherwise plays a negligible role in the electron emission process.

For projectiles whose velocity is larger than the mean velocity of the bound electron, a violent binary collision gives rise to a pronounced maximum which can be observed in the double-differential ionization cross sections, e.g., see the distinct peak at about 55 keV in Fig. 2.1. The location of this maximum is a function of the electron emission angle determined by two-body kinematics. Energy and momentum conservation rules predict that the binary encounter peak has a maximum at an electron energy given by

$$\epsilon_{BE} = 4T \cos^2\theta \quad \text{for} \quad 0 \leq \theta \leq 90°, \tag{2.1}$$

where θ is the electron emission angle and $T = E_p/M_p$ is the projectile energy reduced by the projectile mass M_p (given in units of the electron mass). This reduced quantity is equal to the kinetic energy of an electron with the same velocity as the projectile.

Equation (2.1) applies for an electron initially at rest. In this case, the binary collision peak reduces to a Dirac delta function, which yields a single energy for a given angle of the ejected electron. Such a binary-encounter peak vanishes for angles larger than 90°. As the target electron has an initial velocity distribution due to its binding, the delta function is replaced by a peak profile of nonzero width (Fig. 2.3a). Hence, the effect of the initial electron velocity produces the characteristic binary-encounter peak whose shape is determined by the corresponding Compton profile (Fig. 2.1). The Compton profile is equal to the distribution of the z-momentum component of the bound electron obtained after summing over the other momentum components. The notation "Compton profile" originates from its initial use within the inelastic photon scattering method introduced in the well-known work by Compton (1923).

It should be realized that the initial momentum distribution causes electron ejection at backward angles where the binary-encounter formula (2.1) fails to predict electron emission. The electron emission at backward angles originates primarily from the high-velocity components in the initial velocity distribution. This case corresponds to a relatively slow projectile providing a Coulomb field which may significantly change the momentum direction of the orbiting electron. In fact, this scattering event may be considered as a binary-encounter process where the roles of the projectile ion and the target electron are reversed. In the inverse binary-encounter process the electrons are elastically scattered by the projectile and they may leave the target atom with a velocity as large as their initial velocity (Fig. 2.3b).

It is important to note that the Born approximation describes rather accurately the binary-encounter process which is characterized by a large momentum transfer. This appears to be contradictory since for violent collisions perturbation theory is expected to fail. The resolution of this controversy is due to the remarkable fact that the treatments of the two-body Coulomb problem yield the same result when calculated either in first-order or in higher-order perturbation theory [*Madison* and *Merzbacher* (1975)]. This has already been noted in the discussion of the center concept.

In addition, the quantum-mechanical methods yield the same result as a classical calculation [*Rutherford* (1911)]. From this exceptional agreement of quantum-mechanical and classical methods one is tempted to consider that the binary-encounter approach, as the Born approximation, is a perturbation

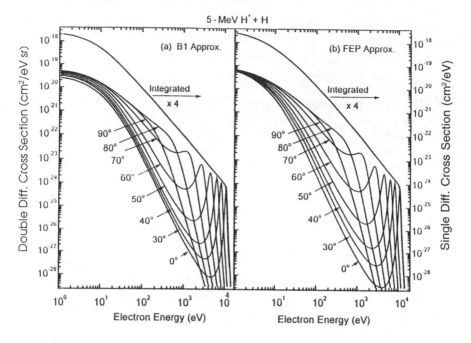

Fig. 2.4a,b. Single differential cross sections (SDCC) and double differential cross sections (DDCS) for electron emission at different angles in 5-MeV H$^+$+He collisions. The data in (**a**) are derived from the Born (B1) approximation formula given by *Landau* and *Lifschitz* (1958). The data in (**b**) are calculated using the free-electron peaking (FEP) model presented in Sect. 3.3.4

theory. This assumption is not valid. It should be realized that the binary-encounter theory includes the effect of the projectile field in all orders. In this sense the binary-encounter theory is superior to the Born approximation when strong projectile-electron interactions are to be treated. Differences between these treatments become noticeable when the phases of the scattering amplitude are probed in the collision. This is the case for dressed projectiles where the results for first- and higher-order perturbation theory no longer coincide [*Reinhold* et al. (1990a)].

Finally, it is noted that the binary-encounter maximum disappears in the single-differential cross section $d\sigma(\epsilon)/d\epsilon$ which is obtained by integrating over the electron emission angle. This is due to the fact that the binary encounter maximum continuously varies in energy as the electron emission angle varies. Therefore, the $d\sigma(\epsilon)/d\epsilon$ curve is a sum of closely-lying binary encounter maxima which produces a structureless, monotonically decreasing function, as illustrated in Fig. 2.4. Hence, apart from low-energy electrons, the single-differential cross section is governed by binary-encounter electron emission. For this reason $d\sigma(\epsilon)/d\epsilon$ drops rapidly to zero for $\epsilon \geq 4\,T$ which corresponds to the BE maximum at 0°, see (2.1).

2.3 Three Bodies: Single-Center Electron Emission

Target-center phenomena in the electron emission become important when the target nucleus significantly influences the outgoing electron. According to Fig. 2.2 the ionization process involves first a binary-encounter event followed by a scattering of the outgoing electron in the Coulomb field of the target nucleus. Figure 2.3d shows an example for the process where a high-energy electron is initially emitted in the forward direction but is backscattered in the field of the target nucleus. This process contributes to the backward emission of electrons.

The influence of the target nucleus increases with decreasing electron energy. In the low-energy limit, the interaction of the target nucleus with the active electron may be stronger than the interaction between the active electron and the projectile. This occurs for glancing collisions and produces the soft-collision (SC) maximum shown in Fig. 2.1. Soft collisions are characterized by small energy and momentum transfers involving weak projectile deflections. In accordance with (2.1), free electrons of near zero energy are primarily ejected near 90°. As mentioned before, the initial electron velocity broadens the corresponding peak structure so that the electrons are emitted over a relatively wide angular range.

However, this broadening effect is generally insufficient to produce the nearly isotropic distribution symmetric around 90° which is observed for the electrons in the low-energy limit [*Manson* et al. (1975)]. The broad angular distribution may be understood from the fact that the target atom becomes a center. When the low-energy electrons are significantly deflected in the field of the target nucleus, as indicated in Fig. 2.3c, they lose their "memory" of the initial 90° emission and a nearly isotropic emission can result. However, classical pictures fail for soft collision electrons, since quantum-mechanical aspects play a dominant role. The *de Broglie* wave length of the soft-collision electrons is comparable with the dimension of the atom so that the wave aspect of these electrons becomes important.

Thus, for a description of the soft-collision electrons it is necessary to use quantum-mechanical methods. The weak projectile interaction, producing the soft-collision electrons, is suitable for the application of perturbation theory. In accordance with the single-center concept used here the initial (bound) state and final (continuum) state of the electron are centered at the target nucleus. Therefore, soft collisions are well described by the first Born approximation. In his fundamental work *Bethe* (1930) has shown by means of the Born approximation that the production of low-energy electrons is governed by dipole-like transitions. The slow-electron production involves a logarithmic projectile energy dependence and the double-differential cross section for emission of low-energy electrons is given by:

$$\frac{d\sigma(\epsilon \rightarrow 0)}{d\epsilon \, d\Omega} = \text{const.} \; \frac{\log T}{T}. \tag{2.2}$$

The derivation of this relation has been given previously in various articles [Bethe (1930), Inokuti (1971), Rudd et al. (1992), McDaniel et al. (1993)] which have also pointed out the similarity of soft collisions and photoionization. Dipole transitions in photoionization transfer negligible momentum to the atom [Berkowitz (1979)]. Likewise, the soft collision peak is characterized by small momentum transfers as noted above.

Finally, a few remarks about total ionization cross sections are in order. Total ionization cross sections are governed by the soft-collision mechanism, since most electrons are produced in glancing collisions. In fact, roughly half of the electron intensity occurs in the low-energy region ranging from zero energy to a value equal to the corresponding binding energy [Rudd et al. (1992)]. Hence, at sufficiently high projectile energies the Born approximation is just as valid in the case of differential cross sections for soft-collision electrons, as it is for the corresponding total ionization cross sections.

2.4 Three Bodies: Two-Center Electron Emission

Two-center effects become important when the outgoing electron is influenced by the superposition of the fields of the target and the projectile nuclei (Fig. 2.3). Two-center electron emission (TCEE) is expected to be significant in the region between the soft collision and the binary-encounter maximum, as shown in Fig. 2.1. It is pointed out that TCEE is observable predominantly in doubly-differential cross sections, since single-differential cross sections are governed by binary-encounter and soft collisions (which are single-center phenomena).

Two-center effects produce an enhancement in the electron emission at forward angles. This finding becomes evident when experimental ionization cross sections are compared with the corresponding results from a one- center theory such as the Born approximation. Enhanced electron emission has been observed in early experiments at forward angles, e.g., 10°, using proton impact at energies of several hundred keV [Rudd et al. (1966a)]. At 0° electron emission increases strongly and the pronounced cusp, due to the process of electron capture to the continuum, occurs [Crooks and Rudd (1970)].

Besides the enhancement of the electron emission at forward angles, other two-center phenomena can be observed. Much effort was devoted to the study of "saddle-point electrons" which are electrons that may be stranded at the saddle between the collision partners. This saddle, formed by the combined Coulomb potentials of the collision partners, has a maximum along the internuclear axis and a minimum perpendicular to it [Meckbach et al. (1986), Olson et al. (1987)].

Furthermore, using fast highly charged projectiles the electron emission was found to be significantly reduced at backward angles [Stolterfoht et al. (1987)]. The enhancement of the TCEE cross sections at forward angles and their reduction at backward angles can be visualized as a result of the attraction of the outgoing electrons by the charged projectile. Evidently, the projectile attracts the ejected electron even if it moves rapidly in a direction opposite to that of the projectile.

In recent years, much effort also was devoted to the verification of two-center effects in spectral structures that are generally attributed to single-center phenomena. For instance, it was realized that soft-collision electrons are influenced by two-center effects when low-energy projectiles are used [*Suárez* et al. (1993a), *Bernardi* et al. (1994)]. Furthermore, the binary-encounter peak was observed to be shifted by two-center effects with respect to the position predicted by two-body kinematics [*Hidmi* et al. (1993)].

The interplay of the electron within the two-center fields of the collision partners can be visualized in the classical double-scattering mechanism postulated by *Thomas* (1927). For instance, consider the ECC mechanism that involves the electron traveling at $0°$ with a velocity equal to the projectile velocity [*Macek* (1991), *Briggs* and *Macek* (1991)]. It follows from (2.1) that an electron with that velocity is produced in a binary-encounter collision at an angle of $60°$, (Fig. 2.3f). Such electrons would rapidly separate from the projectile so that the process of electron capture becomes impossible. However, if it is backscattered by $60°$ in the field of the target nucleus, the electron may travel on a trajectory parallel to the projectile direction. Hence, the electron may be "captured" into continuum states of the projectile as indicated in Fig. 2.3f. Similar target backscattering effects are relevant for the two-center electron emission depicted in Fig. 2.3e. In this case the outgoing electron is not "captured", rather, it remains under the influence of the projectile field after scattering in the target center. Hence, both projectile and target atom play a major role in two-center electron emission.

2.5 Four Bodies: Collisions Involving Projectile Electrons

When the projectile is "dressed", i.e., when it is carrying electrons, the process of target ionization may be significantly affected by the electron incident with the projectile. Hence, additional electron production mechanisms occur as evidenced by modifications of the original peak intensities and/or as new features in the corresponding electron spectra. Figure 2.5 shows examples of results obtained in 30-MeV $O^{q+}+O_2$ collisions for a variety of incident charge states [*Stolterfoht* et al. (1974)]. The peaks associated with the target and the projectile are labeled (T) and (P), respectively. The spectrum for the bare projectile O^{8+} shows similar structures as those seen in Fig. 2.1. One can recognize the soft-collision and the binary-encounter electrons that originate from ionization of the target. In general, the corresponding peak intensities decrease with an increasing number of projectile electrons. The projectile charge-state dependence of target ionization is due to screening effects of the electrons bound at the incident projectile nucleus.

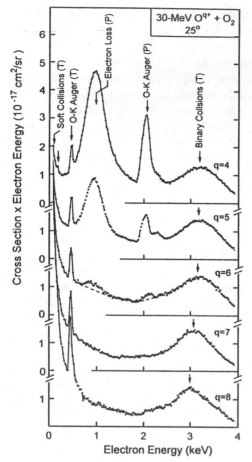

Fig. 2.5. Double-differential cross section for electron emission in collisions of 30-MeV O^{q+} with O_2 observed at an angle of 25°. Note that the cross sections are multiplied by the electron energy. The projectile charge state q is varied from 4 to 8 as indicated. From [*Stolterfoht* et al. (1974)]

Figure 2.5 exhibits peaks due to Auger electron emission from the target, labeled "O-K Auger (T)", following K-shell vacancy production by the incident projectile. Likewise, the projectile ejects K-Auger electrons, labeled "O-K Auger (P)". These electrons are influenced by kinematic effects so that the projectile Auger peaks occur at positions different from that for the target. As noted before, the Auger peaks will not be further discussed here as this topic lies outside the scope of this article. The prominent new feature in the spectra, produced by dressed projectiles, is the electron loss peak, labeled "EL (P)" that originates from ionization of projectile electrons by the interaction with the target atom. In this case, the collision system may be considered as being reversed. As expected, the EL peak increases with increasing number of projectile electrons.

The effects of the projectile electrons on the target ionization were first studied by *Bates* and *Griffing* (1953). Two fundamental roles of the incident electron are indicated in Fig. 2.6 [*Stolterfoht* et al. (1991)]. The first role refers to an incident electron that is assumed to be passive since it remains in its ground state during the interaction with the target electron (Fig. 2.6a). The fact that the state of the projectile electron is unaffected does not indicate a negligible interaction with the target electron, however. Rather, in the collision the target electron may transfer momentum to the projectile electron, but due to the strong coupling to the projectile nucleus the momentum received is further transferred to that nucleus. Hence, the projectile electron acts as a part of the incident nucleus and appears to be a "heavy" particle whose interaction is monoelectronic.

The second role of the incident electron in the Bates-Griffing scattering mechanism refers to the situation where the projectile electron is removed from its ground state (Fig. 2.6b) and plays an active role in the collision. In particular, the removed electron interacts with the target electron on equal grounds. In contrast to the previous case (Fig. 2.6a), the projectile electron acts as a "light" particle. The resulting two-electron event is referred to as a two-center dielectronic process. Dielectronic processes are manifestations of dynamic electron correlation occurring during the collision [*Stolterfoht* (1991)]. The two-center dielectronic process occurs only above a certain

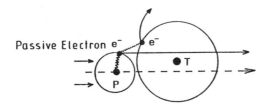

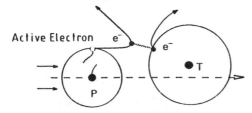

Fig. 2.6. Bates-Griffing scattering mechanism involving a passive electron which remains bound at the projectile during the collision, and an active electron which is liberated and scattered at the target electron. The target electron may get excited or ionized

threshold of the projectile energy since the electron requires a minimum incident velocity equal to the average velocity of the target electron that is ionized, [*Anholt* (1986)]. This means that the dielectronic process requires a reduced projectile energy T larger than the binding energy of the target electron.

The dielectronic process may also cause the ejection of electrons from the projectile. This electron emission contributes to the EL peak that is predominantly produced via interactions with the target nucleus. Hence, the electron data obtained from collisions with dressed projectiles provide detailed information about electron-production mechanisms. In particular, it should be kept in mind that the spectral structures from the target and the projectile can be studied individually when fast projectiles are used in the experiments.

3. Theoretical Work

During the past few decades, various theoretical approaches have been developed to describe electron emission in ion-atom collisions. Each of these approaches uses some approximations that are required to solve the many-body problem associated with the ionization process. In this section, a broad overview of current theoretical methods concerning ionization in ion-atom collisions is given. We shall classify the theoretical methods in terms of the strength of the electronic interactions with the collision partners. This treatment is consistent with our picture of Coulomb centers.

Hence, the theoretical discussion is structured in accordance with the previous section. After the discussion of two-body collisions with bare projectiles, the Born approximation is treated as an example of a target center case. Distorted wave models and the classical trajectory Monte Carlo method are reviewed as examples for two-center interactions. Finally, ionization induced by dressed projectiles are shown to be due to a projectile center when the incident electrons remain passive. Moreover, as the incident electrons become active, dielectronic processes are considered in complex systems of four or more bodies.

Concerning notation, an attempt is made to be as consistent as possible when different formalisms are treated. However, we did not manage to use a unique set of symbols. Different groups have chosen their own characteristic names and symbols which are sometimes difficult to change. Therefore, in particular cases we use such names and symbols. However, within a given subsection a unique set of symbols is applied.

Atomic units are used throughout, unless otherwise stated.

3.1 Two-Body Approximations

We know from classical mechanics that an impulsive force corresponds to a force of large magnitude acting during a small length of time. As stated by *Coleman* (1969): "The analogous situation in the scattering of a particle by an atomic target in a bound state occurs when the duration of the collision is short compared with some time interval characteristic of the target, for example the orbital period of a bound particle. In that case, it may be assumed that, apart from determining the momentum distribution of the states, the target binding forces do not play an important role in the collision".

The above picture is subject to the conditions $v \gg v_e$ (where v and v_e are the collision velocity and the mean orbital velocity of the target electron, respectively) and $k \gg v_e$ (where k is the momentum of the ejected electron measured from the target frame). In the following, we discuss two-body theories that emerge from the impulsive treatment of the target electron.

3.1.1 The Rutherford Formula

The simplest approach to ionization is provided by the Rutherford formula that applies to the scattering of two charged particles in the center-of-mass system. The Rutherford cross section is given by

$$\frac{d\sigma_R}{d\Omega_{cm}} = \frac{Z_1^2 \, Z_2^2}{16 \, \epsilon_{cm}^2 \, \sin^4(\theta_{cm}/2)}, \tag{3.1}$$

where Z_1 and Z_2 are the charges of the two particles, ϵ_{cm} is the kinetic energy in the center of mass, θ_{cm} is the scattering angle of the corresponding reduced particle ejected into the solid angle $d\Omega_{cm} = \sin\theta_{cm} \, d\theta_{cm} \, d\phi_{cm}$. As the Rutherford cross section is formulated within the center-of-mass system, the particles may have significantly different masses. Thus, different versions exist for the Rutherford formula.

For instance, (3.1) covers the interaction between a "free" electron (particle 1) and a heavy particle (particle 2). In this case, $Z_1 = 1$ and $Z_2 = Z$ and the center of mass is practically identical with that of the heavy particle. Hence, one obtains

$$\frac{d\sigma_R}{d\Omega} = \frac{Z^2}{16 \, \epsilon^2 \sin^4(\theta/2)}, \tag{3.2}$$

where ϵ is the energy and θ is the scattering angle of the electron in the rest frame of the heavy particle.

An important feature of the Rutherford cross section for bare particles is that (3.2) implies a singularity at $\theta = 0$. The singularity is removed if the heavy particle is screened by bound electrons. For an exponentially screened Coulomb potential $V(r) = (Z/r) \exp(-r/r_{sc})$ it follows in first-order that

$$\frac{d\sigma_R}{d\Omega} = \frac{Z^2}{[4 \, \epsilon \, \sin^2(\theta/2) + \theta_0^2]^2}, \tag{3.3}$$

where $\theta_0 = (k \, r_{sc})^{-1}$ is the minimum scattering angle determined by the screening length $r_{sc} = Z^{-1/3}$ and the momentum $k = (2\epsilon)^{-1/2}$ of the electron [*Jackson* (1962)].

Finally, we consider the case of an electron initially at rest interacting with a projectile which is incident with the reduced energy T previously defined in (2.2). *Thomson* (1912) presented a version of the Rutherford formula which is differential in the energy transfer Q to the bound electron. This version of the Rutherford formula is given by the simple expression [$Rudd$ et al. (1992)],

$$\frac{d\sigma_R}{d\epsilon} = \frac{\pi Z_p^2}{T Q^2},$$ (3.4)

where Z_p is the nuclear charge of the projectile. Part of the energy transfer is needed to overcome the ionization energy E_b and the remaining energy ϵ is given to the ejected electron so that $Q = E_b + \epsilon$.

Formula (3.4) is of fundamental importance for the theoretical treatment of the ionization process. It indicates that for $\epsilon \gg E_b$ the cross section, differential in electron energy, follows (approximately) an ϵ^{-2} law. Furthermore, no unusual behavior of the cross section is expected as ϵ tends to zero. The latter property is typical for the Coulomb potential where a smooth transition occurs from low-energy continuum states into the associated Rydberg states [$Burgdörfer$ (1984)].

The Rutherford formula is derived under rather different conditions, i.e., it is obtained from classical mechanics as well as from quantum mechanics in both low and high order. Despite its fundamental importance, (3.4) should be used with care. Generally the Rutherford formula noticeably underestimates the electron production cross sections at low electron energies. As pointed out previously, the soft-collision electrons are produced in dipole-type transitions where a quantum treatment is essential.

To enhance the low-energy electrons due to dipole transitions, we modify the original formula (3.4) by introducing an adjustable parameter c in the denominator:

$$\frac{d\sigma}{d\epsilon} = \frac{\pi Z_p^2}{T (c E_b + \epsilon)^2}.$$ (3.5)

With the c parameter one may govern the cross section values for small electron energies. To achieve agreement with the Born approximation [$Bethe$ (1930)], we incorporated the logarithmic projectile velocity dependence mentioned in Sect. 2.3. Hence, in accordance with (2.2) we introduce the semiempirical expression

$$c = \left[\ln\left(\frac{2T}{E_b}\right)\right]^{-1/2}.$$ (3.6)

For high projectile energies, relevant in this article, this expression yields $c \approx 0.5$ varying only slightly with T and E_b. It will be shown in later sections that (3.5) in conjunction with (3.6) works remarkably well for both soft-collision and binary-encounter electrons. It predicts interesting features for these types of electrons: The single differential cross sections of the binary-encounter electrons are independent of target properties, whereas the corresponding cross sections of the soft-collision electrons are governed by the ionization energy E_b.

Finally, (3.5) is integrated to obtain the total cross section for electron emission. The integration is performed analytically yielding the simple expression

$$\sigma_{tot} = \frac{Z_p^2 \, \pi}{T \, E_b} \left(\ln \frac{2T}{E_b} \right)^{1/2}. \tag{3.7}$$

This formula may serve as a convenient tool for roughly estimating total ionization cross sections. We note that (3.7) contains a useful scaling law. It yields a universal curve when the quantity $E_b^2 \, \sigma_{tot}$ is plotted versus T/E_b. The same scaling rule has been extracted from the binary encounter theory by *Garcia* (1970).

3.1.2 Elastic Scattering Model

In the past few years, much effort has been devoted to analyzing the profile of the binary encounter peak. *Bell* et al. (1983) introduced a theoretical treatment based on the Compton profile to describe the binary-encounter mechanism. This treatment will be discussed further below in conjunction with the Born approximation. More recently, *Lee* et al. (1990) put forward a similar model to interpret extensive electron measurements at an observation angle of 0°. The model involves an impulse approximation which adopts a free incident electron, liberated from the target before the collision, hence, *Lee* et al. (1990) refer to their theoretical model as the impulse approximation (IA). To avoid confusion with this notation in other theoretical contexts, the present model is also referred to as the elastic scattering model [*Richard* (1996)]. It should be kept in mind that this model, similar to the binary-encounter theory, involves to a single projectile-center.

Basic Model. The basic model assumption is that target electrons form a parallel beam incident on the projectile. Thus, the evaluation of the model begins in the projectile frame of reference and a projectile-to-laboratory frame transformation is applied to the final result. Hence, the double differential cross section for electron emission is evaluated as [*Brandt* et al. (1983), *Lee* et al. (1990), *Zouros* et al. (1994)]

$$\frac{d\sigma}{d\Omega \, d\epsilon} = \sqrt{\frac{\epsilon}{\epsilon'}} \int \frac{d\sigma'(\epsilon', \theta')}{d\Omega'} \, f(\boldsymbol{p}) \, \frac{dp_z}{d\epsilon'} \, d^2 \boldsymbol{p}_\perp, \tag{3.8}$$

where $d\sigma'(\epsilon', \theta')/d\Omega'$ is the cross section for an electron of energy ϵ', elastically scattered through the angle θ' in the projectile center and $f(\boldsymbol{p}) = |\tilde{\varphi}(\boldsymbol{p})|^2$ is the momentum distribution of the target electron obtained from its wave function $\tilde{\varphi}(\boldsymbol{p})$ in momentum space. The integration is performed with respect to the initial electron momentum $\boldsymbol{p}_\perp$ perpendicular to the incident beam direction. This latter direction is taken as the z axis so that

$d^2\boldsymbol{p}_\perp = dp_x\, dp_y$. The primed quantities refer to the projectile frame of reference. The factor $\sqrt{\epsilon/\epsilon'}$ originates from the projectile-to- laboratory frame transformation outlined in Appendix B.

The incident energy of the electron is obtained from the sum of the electron momentum $\boldsymbol{p}$ and the incident velocity $\boldsymbol{v}$

$$\epsilon' = \frac{1}{2}(\boldsymbol{v}+\boldsymbol{p})^2 - E_b = \frac{1}{2}\left(v^2 + 2vp_z + p_z^2 + p_\perp^2\right) - E_b$$

$$\approx \left(\sqrt{T}+\frac{p_z}{\sqrt{2}}\right)^2 - E_b, \tag{3.9}$$

where $T = v^2/2$ is the reduced projectile energy. It is noted that the (positive) ionization energy E_b is subtracted from the kinetic energy of the ejected electron (*Lee* et al., 1990). The subtraction implies that the energy required for ionization is uniquely taken from the electron. This may lead to an unphysical situation since for $\boldsymbol{p} = -\boldsymbol{v}$ the electron energy ϵ' becomes negative. Alternatively, one may assume that the ionization energy is provided by the projectile so that the subtraction of E_b is suppressed.

In the final expression of (3.9), the term $\boldsymbol{p}_\perp^2$ is neglected in comparison with v^2. This is the characteristic approximation of the present model. It is well justified for sufficiently large values of v^2 (or T) relevant for fast projectiles. In this case, the incident electron energy ϵ' becomes independent of $\boldsymbol{p}_\perp$ so that (3.8) is substantially simplified:

$$\frac{d\sigma}{d\Omega\, d\epsilon} = \sqrt{\frac{\epsilon}{\epsilon'}}\frac{d\sigma'(\epsilon',\theta')}{d\Omega'}\frac{dp_z}{d\epsilon'}\int |\tilde{\varphi}(\boldsymbol{p})|^2\, d^2\boldsymbol{p}_\perp. \tag{3.10}$$

The dependence of the scattering angle on the transverse momentum will be discussed further below. From (3.9) it can be shown that $dp_z/d\epsilon' = \left(\sqrt{2T}+p_z\right)^{-1}$. Finally, the double differential cross section is obtained as

$$\frac{d\sigma}{d\Omega\, d\epsilon} = \sqrt{\frac{\epsilon}{\epsilon'}}\frac{d\sigma'(\epsilon',\theta')}{d\Omega'}\frac{J(p_z)}{\sqrt{2T}+p_z}, \tag{3.11}$$

where $J(p_z) = \int |\tilde{\varphi}(\boldsymbol{p})|^2\, d^2\boldsymbol{p}_\perp$ is the Compton profile of a scattered electron initially bound to the target atom.

To evaluate the double differential cross section, the z component of the initial electron momentum p_z is expressed in terms of the incident electron energy ϵ' by inverting (3.9) with neglect of $\boldsymbol{p}_\perp^2/2$:

$$p_z = \sqrt{2}\left(\sqrt{\epsilon'+E_b}-\sqrt{T}\right). \tag{3.12}$$

The expressions for transforming the quantities ϵ' and Ω' from the projectile to the laboratory frame of reference are given in Appendix B.

The final expression (3.11) of the double differential cross section is noted to be of remarkable simplicity as it is evaluated without numerical integration.

For hydrogenic wave functions, the Compton profile can be evaluated in closed form. For instance, the hydrogenic Compton profile for the 1s electron is given by

$$J_{1s}(p_z) = \frac{8}{3\,\pi} \frac{Z_t^5}{\left(Z_t^2 + p_z^2\right)^3},$$ (3.13)

which is normalized to unity according to $\int J_{1s}(p_z)\, dp_z = 1$. For many-electron systems more effort is required to evaluate the Compton profile. However, for most systems tabulated data for Compton profiles are available in the literature by *Biggs* et al. (1975).

   For an evaluation of (3.11) it remains to determine the single differential cross section $d\sigma'/d\Omega'$ for elastic electron scattering. For bare projectiles, $d\sigma'/d\Omega'$ is equal to the Rutherford cross section given by (3.2). This case will be treated in the experimental Sect. 5.1 devoted to the case of bare projectiles. When the incident ion carries electrons the projectile potential is no longer Coulombic. Thus, more elaborate work is required to determine the elastic scattering cross section. For dressed projectiles we consider first- and higher-order effects of the mean field of the passive electrons. These cases will be treated in the theoretical Sect. 3.7.1 and in the experimental Sect. 6.3 devoted to dressed projectiles.

Extended Model. We now consider an approximate extension of the elastic scattering model. The characteristic approximation of the basic model, namely that target electrons form a parallel beam incident on the projectile, neglects the transverse momentum of the electrons. In specific cases, this approximation causes severe problems. It is recalled that the binary encounter electrons, observed at a laboratory observation angle of 0°, are scattered by 180° in the projectile frame. However, electrons ejected with energies less than those of the cusp electrons are scattered by 0° in the projectile frame. It is well known that the Rutherford cross section diverges at 0° so that a singularity occurs at this angle for small electron energies. The same is true for the laboratory observation angle of 180°.

   To remove these singularities the transverse momentum of the incident target electrons has to be taken into account. In this case the electrons are incident with the direction $\hat{p}'$ of the initial momentum $p' = v + p$ in the projectile frame and scattered into the direction $\hat{k}'$ of $k' = v + k$. The important consequence is that the elastic scattering cross section becomes dependent on the incident momentum direction:

$$\frac{d\sigma}{d\Omega\, d\epsilon} = \sqrt{\frac{\epsilon}{\epsilon'}} \int \frac{d\sigma'(\epsilon', \theta', \hat{p}')}{d\Omega'}\, f(p)\, \frac{dp_z}{d\epsilon'}\, d^2 p_\perp.$$ (3.14)

In fact, the elastic scattering cross section depends on the relative angle α' between $\hat{p}' = (\theta_p', \varphi_p')$ and $\hat{k}' = (\theta', 0)$ that can be determined using $\cos\alpha' = \cos\theta' \cos\theta_p' + \sin\theta' \sin\theta_p' \cos\varphi_p$.

To maintain the analytic formula (3.11), we treat the transverse momentum in an approximation. A constant value is chosen for the incident beam direction. This value is determined by the two major effects producing transverse momenta. We take into account, first, the initial transverse momenta $p_\perp$ and, second, the minimum momentum transfer K_m that is required to ionize the target electron. Both momenta produce an angular divergence of the beam of electrons. This spreading is taken into account by replacing the observation angle in (3.11) by an increased scattering angle, i.e., $\theta' \rightarrow \theta'_s$.

In an approximation we determine the scattering angle θ'_s by adding the observation angle θ' and angular beam spread θ'_b in quadrature

$$\theta'^2_s = \theta'^2 + \theta'^2_b. \tag{3.15}$$

The spreading angle is estimated as

$$\tan \theta_b = \sqrt{\frac{\frac{1}{2}(p_\perp^2 + K_m^2)}{v_e'^2}}, \tag{3.16}$$

where $v_e' = \sqrt{2\epsilon'}$ is the incident electron velocity in the projectile frame of reference. The contributions associated with the transverse momentum and the minimum momentum transfer are again added in quadrature. The transverse momentum is estimated from the corresponding ionization energy of the target electron $p_\perp \approx p/\sqrt{2} \approx \sqrt{E_b}$ and we set $K_m = (E_b + \epsilon)/v$ (Sect. 3.4.2).

We would not expect that the present treatment completely cancels the problems inherent in the simple formula (3.11). Although the singularity is removed, care should be taken with forward and backward angles. However, these limiting cases do not play a significant role for cross sections integrated over the solid angle.

It is interesting to add that the present formalism of the elastic scattering model reproduces a cusp-shaped peak. Its shape is even found to be asymmetric, with a significant enhancement of the low-energy wing, which follows from the fact that $0°$ scattering is more probable than $180°$ scattering. At present, we would consider this finding as an artifact of the model. A theoretical approach involving a single projectile center (e.g., the binary encounter theory) does not account for the ECC mechanism. To produce the ECC cusp, the influence of the target center cannot be neglected during the collision. The elastic scattering picture may help for a qualitative understanding of the ECC process, but is not suitable for quantitative predictions. For very small electron velocities in the projectile reference frame, the initial angular spread of the electrons has to be adequately taken into account. This is done in the binary-encounter theory treated next.

3.1.3 Binary Encounter Approximation

The binary-encounter approximation (BEA) introduced by *Gryzinski* (1959) and further developed by Vriens and coworkers [*Vriens* and *Bonsen* (1968),

Banks et al. (1969), *Vriens* (1969), *Bonsen* and *Vriens* (1970, 1971)] is a typical impulsive approximation. In this model the target atomic electron is assumed to be free and, hence, it interacts with the incident projectile independent of the target atom. However, the BEA accounts for the initial velocity of the electrons bound to the target atom.

Starting from the Rutherford formula (3.1) and after lengthy but straightforward algebra [*Vriens* (1969), *Bonsen* and *Vriens* (1970)], the double differential cross section for ejection of a "bound" electron between the scattering angles θ and $\theta + d\theta$ and with an energy between $E + \epsilon_i$ and $E + \epsilon_i + dE$ (with ϵ_i the binding energy of the electron to be ejected) is given by

$$\sigma(E, \theta, v_2) = 2\pi \sin \theta \sigma(E, \boldsymbol{v}_2', v_2) = 2\pi \sin \theta \frac{Z_P^2 v_1 v_2'}{2 v_2 E^3}$$

$$\times \left\{ v_2'^2 \sin^2 \theta - E \left[1 - \frac{v_2'}{v_1}(1 + \frac{1}{m_1}) \cos \theta + \frac{v_2'^2}{m_1 v_1^2} \right] \right\} |v_1 - v_2'|^{-3}.$$

$$(3.17)$$

In (3.17), $\boldsymbol{v}_1$ ($\boldsymbol{v}_2$) and $\boldsymbol{v}_1'$ ($\boldsymbol{v}_2'$) are the initial and final velocities of the particle 1 (2), and m_1 is its mass. The mass of the electron has been taken as $m_2 = 1$. The angle θ is formed between $\boldsymbol{v}_2'$ and $\boldsymbol{v}_1$ and E is the energy transfer: $E = v_2'^2/2 - v_2^2/2$.

The difference between $\sigma(E, \theta, v_2)$ and $\sigma(E, \hat{\boldsymbol{v}}_2', v_2)$ lies in the fact that the first differential cross section is given as a function of the polar angle θ and integrated over the azimuthal angle subtended by $\hat{\boldsymbol{v}}_2'$ and the second differential cross section is given per unit of solid angle subtended by $\hat{\boldsymbol{v}}_2'$. To obtain $\sigma(E, \theta, v_2)$ it has been assumed that the target electrons have an isotropic velocity distribution and integrations have been made over the angle formed between $\boldsymbol{v}_2$ and $\boldsymbol{v}_1$ and over the azimuthal angle of $\boldsymbol{v}_2$ with respect to $\boldsymbol{v}_1$.

To allow for the comparison with the experiments, $\sigma(E, \hat{\boldsymbol{v}}_2', v_2)$ is integrated over the velocity distribution $f(v_2)$ of the initial bound state, so that

$$\frac{d\sigma}{d\epsilon \, d\Omega} = \int_{v_{2\,\text{min}}}^{v_{2\,\text{max}}} \sigma(E, \hat{\boldsymbol{k}}, v_2) \, f(v_2) \, dv_2, \tag{3.18}$$

where $\hat{\boldsymbol{k}} = \hat{\boldsymbol{v}}_2'$ and $k^2 = 2(E + \epsilon_i)$ [*Vriens* (1970)]. We have defined the final electron energy as $\epsilon = k^2/2$. The integration limits of (3.18) are obtained from the conditions imposed by the laws of conservation of energy and momentum [*Bonsen* and *Vriens* (1970)]. In (3.18),

$$\int_0^\infty f(v_2) \, dv_2 = 1, \tag{3.19}$$

in order to normalize the initial velocity distribution.

Calculations of $d\sigma/d\epsilon \, d\Omega$ have been compared with experimental data by *Rudd* et al. (1966a) for impact of 100-keV and 300-keV protons on He and

H_2 targets. Good agreement is obtained in the region of the binary encounter peak as k increases. However, a poor description of the experiments is found at small and large electron scattering angles and at small k.

In order to improve the BEA in $H^+ + He$ collisions, *Bonsen* and *Vriens* (1971) took into account the polarization of a helium-like target atom by the proton using the polarized orbital method of *Temkin* (1959). If r_1 indicates the position of the proton and r_2 and r_3 the positions of the atomic electrons with respect to the target nucleus, the initial distorted bound state was chosen as

$$\Psi_i(r_2, r_3) = \prod_{j=2}^{3} [\varphi_i(r_j) + \varphi_i^{pol}(r_j, r_1)] \tag{3.20}$$

with

$$\varphi_i(r_j) + \varphi_i^{pol}(r_j, r_1)$$
$$= \left(\frac{Z^3}{\pi}\right)^{1/2} e^{-Zr_j} \left\{ 1 + \frac{\varepsilon(r_1, r_j)}{(Zr_1)^2} \left(r_j + \frac{Zr_j^2}{2} \right) \cos\theta_{1j} \right\}, \tag{3.21}$$

where Z is an effective target nuclear charge corresponding to a hydrogenic approximation of the target. In (3.21), θ_{1j} is the angle between r_1 and r_j and $\varepsilon(r_1, r_j)$ is equal to unity for $r_1 > r_j$ and equal to zero for $r_1 < r_j$. This model corresponds to considering only the first term of the multiple expansion of the interaction potential between the proton and the complete target; it is a dipole polarization potential for large enough distances $r_1 > r_j$. This interaction is taken as zero for $r_1 < r_j$. Transforming the wave function (3.21) to velocity space, the velocity distribution for each atomic electron is obtained. Thus, double differential cross sections are calculated as in the previous case.

This combined polarized-orbital binary-encounter theory improves the agreement with experimental double differential cross sections for impact of 100-keV to 300-keV protons on He, using data at a fixed electron scattering angle $\theta = 10°$. However, the experimental hump [*Rudd* et al. (1966a)], associated with two-center effects cannot be described by this simple one-center theory.

3.2 Classical Trajectory Monte Carlo Method

The classical trajectory Monte Carlo method (CTMC) was introduced by *Abrines* and *Percival* (1966a,b) to study ion-atom collisions. All previous classical theories used two basic assumptions: (i) The particles obey the Newtonian laws during the collision and, (ii) the many-particle collision is approximated by two-body collisions for each atomic electron. In the model proposed by Abrines and Percival, only approximation (i) is used, so that a more accurate classical treatment of collisions involving three particles is developed. The model can be understood in terms of classical mechanics including statistical aspects.

CTMC is a numerical method used to calculate the time evolution of a classical distribution $f(\boldsymbol{x}, \boldsymbol{p}, t)$ in phase space. We consider a collision system composed by two frozen cores (projectile and target cores) of masses M_P and M_T and one active electron initially bound to the target.

For a given set of initial conditions in the phase space, the dynamics of the system is determined by the classical Hamilton equations:

$$\dot{x}_j = \frac{\partial H}{\partial p_j} = \frac{p_j}{\mu_T}, \quad \dot{p}_j = -\frac{\partial H}{\partial x_j}, \quad j = 1, 2, 3 \tag{3.22}$$

$$\dot{R}_{T_j} = \frac{\partial H}{\partial P_j} = \frac{P_j}{\nu_T}, \quad \dot{P}_j = -\frac{\partial H}{\partial R_{T_j}}, \quad j = 1, 2, 3 \tag{3.23}$$

where (x_j, p_j) and (R_{T_j}, P_j) are phase space coordinates, and the Hamiltonian H is given by

$$H = \frac{P^2}{2\nu_T} + \frac{p^2}{2\mu_T} + V_T(\boldsymbol{x}) + V_P(\boldsymbol{s}) + W(\boldsymbol{R}), \tag{3.24}$$

where $\nu_T = M_P(M_T + 1)/(M_T + M_P + 1)$, and $\mu_T = M_T/(M_T + 1)$. Also V_T, V_P, and W are the electron-target core, electron-projectile, and projectile-target core interactions, respectively. We note that $\boldsymbol{R}_T$ is the position of the projectile with respect to the center of mass of the electron-target core sub-system and $\boldsymbol{x}$ $(\boldsymbol{s})$ is the position of the electron with respect to the target (projectile) nucleus and $\boldsymbol{R}$ the internuclear vector.

For simplicity, the projectile and the target core can be assumed to move rectilinearly corresponding to the straight line version of the impact parameter approximation. This approximation does not affect the reactions of our interest and reduces the problem to the solution of the Hamilton equations (3.22), being now

$$H = \frac{p^2}{2\mu_T} + V(\boldsymbol{x}, t) = \frac{p^2}{2\mu_T} + V_T(\boldsymbol{x}) + V_P(\boldsymbol{s}). \tag{3.25}$$

An important property is that each element of volume in phase space evolves independently, decoupled from the other elements.

The classical distribution in phase space can also be described by the Liouville equation [*Huang* (1963), *Reinhold* (1987), *Reinhold* and *Falcón* (1988a)]:

$$\frac{\partial f}{\partial t} = L(t)f, \tag{3.26}$$

where the Liouville operator L(t) is given by

$$L(t) = \nabla_x V(\boldsymbol{x}, t) \cdot \nabla_p - \frac{\boldsymbol{p}}{\mu_T} \cdot \nabla_x. \tag{3.27}$$

Evaluating (3.26), we obtain [*Reinhold* and *Falcón* (1988a)]

$$\frac{\partial \rho}{\partial t} + \mathrm{div}(\rho \, \boldsymbol{v}) = 0, \tag{3.28}$$

$$\mu_T \frac{dv}{dt} = -\nabla_x V(x,t) - \frac{1}{\rho} \nabla_x \tilde{P}, \qquad (3.29)$$

where the density ρ, the mean velocity v and the strength tensor $\tilde{P}$ are defined as:

$$\rho(x,t) = \int d^3p \; f(x,p,t), \qquad (3.30)$$

$$\mu_T \, \rho(x,t) \, v(x,t) = \int d^3p \; p \, f(x,p,t), \qquad (3.31)$$

and

$$\mu_T \, P_{ij} = \int d^3p \; p_i \, p_j \, f(x,p,t) - \mu_T^2 \, \rho \, v_i \, v_j. \qquad (3.32)$$

It should be realized that (3.28,29) are similar to those obtained from a quantum mechanical treatment of the collision [*Reinhold* and *Falcón* (1988a)]. In this case, the evolution of the active electron is given by the Schrödinger equation:

$$\left[-\frac{1}{2\mu_T} \nabla_x^2 + V(x,t) - i\frac{\partial}{\partial t} \right] \Psi_i^+(x,t) = 0, \qquad (3.33)$$

where $\Psi_i^+(x,t)$ is the exact wave function which satisfies correct outgoing conditions, and at the time $(-T)$ before the collision verifies the limit

$$\lim_{t \to -T} \Psi_i^+(x,t) = \varphi_i(x) \, e^{-i\epsilon_i t} \qquad (3.34)$$

with $\varphi_i(x)$ the initial orbital wave function and ϵ_i the corresponding orbital energy. The wave function $\Psi_i^+(x,t)$ can be written in the form

$$\Psi_i^+(x,t) = \sqrt{\rho(x,t)} \, \exp[iS(x,t)] \qquad (3.35)$$

so that with (3.33) we obtain

$$\frac{\partial \rho}{\partial t} + \text{div}(\rho \, v) = 0, \qquad (3.36)$$

$$\mu_T \frac{dv}{dt} = -\nabla_x V(x,t) - \nabla_x V_q(x,t), \qquad (3.37)$$

where we have used the definitions

$$v(x,t) = \frac{\nabla_x S}{\mu_T}, \qquad (3.38)$$

$$V_q(x,t) = -\frac{1}{2\mu_T} \frac{\nabla_x^2 \sqrt{\rho}}{\rho}, \qquad (3.39)$$

and d/dt denotes the operator $(\partial/\partial t + v \cdot \nabla_x)$.

Now, comparing (3.28,29) with (3.36,37), we can relate the quantum and classical descriptions of the collision, identifying the quantum mechanical potential V_q and the classical strength tensor $\tilde{P}$ through the equation [*Reinhold and Falcón* (1988a)]:

$$\nabla_x V_q = \frac{1}{\rho} \nabla_x \tilde{P}. \tag{3.40}$$

The initial classical distribution $f(x, p, t = -T) = f_i(x, p)$ should reproduce the momentum and position distributions of the initial quantum state:

$$\rho_i(x) = \int d^3p \ f_i(x, p), \tag{3.41}$$

$$\rho_i(p) = \int d^3x \ f_i(x, p). \tag{3.42}$$

The CTMC method deals with statistical averages over an ensemble of possible initial conditions. In practice a sample of a finite number of particles of the phase space describing the initial distribution is considered. The momentum equations are integrated separately for each of these particles.

In order to indicate the way to implement the initial conditions in the CTMC method we will strictly follow the method proposed by *Reinhold* and *Falcón* (1986). The chosen ensemble consists of a monoenergetic uniform beam, moving in the positive direction parallel to the impact velocity v_p colliding with atoms, all of which have the same initial energy ϵ_i [*Abrines and Percival* (1966a)]. The results should be independent of the particular choice of the initial position of the projectile with respect to the target. If $b_{\max}$ is the impact parameter for which the studied reaction is negligible, the uniform flux of incident particles is reproduced choosing b^2 uniformly in the interval $[0, b_{\max}^2]$.

The initial electronic state is obtained from a microcanonical distribution corresponding to a spherically symmetric Hamiltonian, so that

$$f_i(x, p) = k \ \delta \left[\epsilon_i - \frac{p^2}{2\mu_T} - V_T(x) \right] \tag{3.43}$$

with k a normalization constant. The electronic coordinate follows the condition

$$\frac{p^2}{2\mu_T} = \epsilon_i - V_T(x) > 0. \tag{3.44}$$

Assuming that $\epsilon_i - V_T(x) = 0$ has only one root x_0, x is confined in the interval $0 < x < x_0$.

A set of uniformly distributed pseudo-random numbers are used to specify completely the initial state. They are related to the variables (x, p) by the transformations:

$$\begin{aligned} x_1 &= x \, (1 - \nu_x^2)^{1/2} \cos \varphi_x, \\ x_2 &= x \, (1 - \nu_x^2)^{1/2} \sin \varphi_x, \\ x_3 &= x \, \nu_x; \end{aligned} \tag{3.45}$$

$$p_1 = \{2\mu_T[E - V_T(x)]\}^{1/2} \left(1 - \nu_P^2\right)^{1/2} \cos\varphi_P,$$
$$p_2 = \{2\mu_T[E - V_T(x)]\}^{1/2} \left(1 - \nu_P^2\right)^{1/2} \sin\varphi_P, \qquad (3.46)$$
$$p_3 = \{2\mu_T[E - V_T(x)]\}^{1/2}.$$

Also, we define the quantity

$$w(x) = \int_0^x dx'\, \mu_T\, x'^{\,2}\, \{2\mu_T\, [\epsilon_i - V_T(x')]\}. \qquad (3.47)$$

The new quantities range in the intervals

$$\begin{array}{rcl}
E & \in & (-\infty, 0), \quad w \ \in \ [0, w(x_0)], \\
\nu_x, \nu_p & \in & [-1, 1], \quad \varphi_x, \varphi_p \ \in \ [0, 2\pi].
\end{array} \qquad (3.48)$$

The microcanonical distribution, as a function of these new variables, results in:

$$f_i(E, w, \nu_x, \nu_p, \varphi_x, \varphi_p) = k\delta(E - \epsilon_i) \qquad (3.49)$$

which is independent of $(w, \nu_x, \nu_p, \varphi_x, \varphi_p)$. Then, choosing these variables at random, the corresponding initial conditions $(\boldsymbol{x}, \boldsymbol{p})$ are obtained. Different choices of the initial classical distribution can be found, for example, in *Eichenauer* et al. (1981), *Hardie* and *Olson* (1983), and *Cohen* (1985). The dynamics of each particle of the phase space is then evaluated by numerically integrating the corresponding Hamilton equations. Each of the initial conditions generates an independent "history". This classical microcanonical distribution gives a good description of the initial momentum distribution but a poor description of the initial position distribution.

In order to determine the cross sections for different reactions, we define the target and projectile binding energies of the phase-space particles:

$$E_T = \frac{p^2}{2\mu_T} + V_T(\boldsymbol{x}) \qquad (3.50)$$

and

$$E_P = \frac{(\boldsymbol{p} - \mu_T \boldsymbol{v}_P)^2}{2\mu_T} + V_P(\boldsymbol{x} - \boldsymbol{b} - \boldsymbol{v}_P t), \qquad (3.51)$$

where $\boldsymbol{v}_P > 0$ is the collision velocity. At the end of the collision, one of the following situations is obtained:

(i) $E_T > 0$ and $E_P > 0$, corresponds to electron ionization;
(ii) $E_T > 0$ and $E_P < 0$, corresponds to electron capture;
(iii) $E_T < 0$ and $E_P > 0$, the electron remains bound to the target.

For a fixed impact parameter, the transition probabilities can be obtained using the expression

$$P_{i,c,e}(\boldsymbol{b}) = \frac{N_{i,c,e}(\boldsymbol{b})}{N(\boldsymbol{b})}, \qquad (3.52)$$

where N_i, N_c and N_e are the number of particles corresponding to the reaction, (i), (ii), and (iii), respectively, and N is the total number of histories corresponding to the impact parameter b. The associated total cross section can be found by integration of (3.52) over b.

Another method to calculate the ionization and capture total cross section can be obtained by considering the condition of the uniform projectile beam. We thus choose b^2 uniformly and at random in the interval $[0, b_{max}^2]$, so that

$$\sigma_{i,c} = \frac{N_{i,c}}{N} \pi \, b_{max}^2, \tag{3.53}$$

where now N_i and N_c are the number of particles ionized or captured, according to the conditions (i) and (ii), respectively, and N is the total number of possible histories.

Single-differential cross sections for ionization, as a function of the angle or of the energy of the ejected electron, can be calculated as

$$\frac{d\sigma}{d\Omega} = \frac{N_\Omega}{N \Delta\Omega} \pi \, b_{max}^2, \tag{3.54}$$

$$\frac{d\sigma}{d\epsilon} = \frac{N_\epsilon}{N \Delta\epsilon} \pi \, b_{max}^2, \tag{3.55}$$

where N_Ω and N_ϵ are the number of particles detected in the intervals $[\Omega - \Delta\Omega/2, \Omega + \Delta\Omega/2]$ and $[\epsilon - \Delta\epsilon/2, \epsilon + \Delta\epsilon/2]$, respectively. In a similar way we can obtain $d\sigma/d\epsilon\, d\Omega$.

The CTMC model applies for impact velocities comparable to the initial orbital velocity and projectile charges of the order or greater than the initial mean orbital velocity. It must also be remarked that in this model the active electron is considered as moving in the presence of the target and projectile fields and so, in this sense, it is a two-center approximation. It has been found that the CTMC model gives an adequate description of experimental data for target ionization by impact of light particles [*Schultz* and *Reinhold* (1990), *Schultz* et al. (1992)] and heavy particles [*Olson* (1980, 1983, 1986), *Olson* and *Salop* (1977), *Olson* et al. (1993), *Reinhold* and *Falcón* (1988b), *Reinhold* and *Schultz* (1989), *Schultz* et al. (1991), *Schultz* and *Olson* (1991), *Reinhold* et al. (1991), *Schultz* et al. (1992), *Schultz* et al. (1996), *Kerby* et al. (1995)].

The correspondence between classical and quantum descriptions of ionization in fast ion-atom collisions has been studied in detail by *Reinhold* and *Burgdörfer* (1993). They have shown that this correspondence exists for large momentum transfer, but for small momentum transfer ionization is classically suppressed. Furthermore, the CTMC approximation fails to describe large impact parameter reactions and also small impact parameter ionization at high collision energies. All these processes, for which the classical model breaks down, should be described using quantum mechanical methods [*Boesten* and *Bonsen* (1975), *Boesten* et al. (1975), *Reinhold* and *Burgdörfer* (1993)].

3.3 Semiclassical Approximation

The semiclassical approximation (SCA) is a useful tool to describe electron emission in ion-atom collisions, [*Bang* and *Hansteen* (1957), *McDowell* and *Coleman*, (1970)]. This approximation, which treats the motion of the collision partners classically, is expected to work well at the high projectile energies relevant for this work. The semiclassical approximation is applied in the framework of the coupled channel method which will be treated in a later section. Here, we shall use time-dependent perturbation theory.

Today, it is commonly assumed that the "semiclassical approximation" implies first-order perturbation theory [*Hansteen* (1975), *Kocbach* and *Briggs* (1984), *Trautmann* et al. (1985)]. For many purposes this semiclassical approximation is as accurate as the Born approximation. Nevertheless, the theoretical work using the SCA for the evaluation of differential electron emission cross sections is limited. In the following, we present the basic formalism of the semiclassical approximation, followed by a partial wave expansion. As a useful application of the partial wave expansion, the dipole approximation is discussed.

3.3.1 Basic Formalisms

We assume that the target subsystem, described by the Hamiltonian H°, is perturbed by a time-dependent projectile interaction $V(t)$ producing the time-dependent Hamiltonian

$$H = H^\circ + V(t). \tag{3.56}$$

The first-order amplitude for transitions from the initial state i to the final state f is obtained as

$$a_k = -\mathrm{i} \int_{-\infty}^{\infty} \mathrm{e}^{-\mathrm{i}\Delta E t} \, V_{if}(t) \, dt, \tag{3.57}$$

where $V_{if} = \langle f \,|\, V \,|\, i \rangle$ is the corresponding transition matrix element and $\Delta E = E_i - E_f$ is the energy difference between the initial and final state. The electronic perturbation $V = Z_p / |\boldsymbol{R} - \boldsymbol{r}|$ is evaluated from the position vectors $\boldsymbol{R}$ and $\boldsymbol{r}$ of the incident ion and the active target electron, respectively.

In the semiclassical approximation the internuclear motion is treated classically, so that the coordinate of the projectile is a function of the time t. In the high-energy approximation of a straight-line trajectory one obtains the following expression for the internuclear vector:

$$\boldsymbol{R} = \boldsymbol{b} + \boldsymbol{v}\, t, \tag{3.58}$$

where $\boldsymbol{b}$ is the impact parameter and $\boldsymbol{v}$ is the projectile velocity. Hence, the coordinate of the projectile along the z axis parallel to the incident beam direction is given by

$$z = v\, t. \tag{3.59}$$

Replacing the time t by z/v in accordance with (3.59) one obtains

$$a_k = -\frac{i}{v} \int_{-\infty}^{\infty} e^{-iK_m z} \, V_{if}(z/v) \, dz, \tag{3.60}$$

where $K_m = \Delta E/v$ is the minimum momentum transfer.

3.3.2 Partial Wave Expansion

To calculate the transition amplitude, it is useful to perform a partial wave expansion. The initial bound state is factored as

$$\varphi_{n\ell m}(\boldsymbol{r}) = R_{nl}(r) \, Y_l^m(\hat{\boldsymbol{r}}). \tag{3.61}$$

In the following we shall restrict the analysis to s states:

$$\varphi_{ns}(\boldsymbol{r}) = (4\pi)^{-1/2} \, R_{ns}(r). \tag{3.62}$$

The final continuum state is developed into partial waves

$$\varphi_k(\boldsymbol{r}) = \frac{4\pi}{(2\pi)^{2/3}} \sum_{lm} i^l \, e^{i\sigma_l} \, Y_l^{m*}(\hat{\boldsymbol{k}}) \, f_{kl}(r) \, Y_l^m(\hat{\boldsymbol{r}}), \tag{3.63}$$

where σ_l is the phase shift. The continuum wave function $\varphi_k(r)$ and the corresponding radial wave function $f_{kl}(r)$ are normalized with respect to unit momentum. The latter function may be obtained from a wave function $f_{\epsilon l}$ normalized to unit energy as $f_{kl} = \sqrt{\pi/2k} \, f_{\epsilon l}$. The radial wave function depends on the target potential. For instance, for Coulomb wave functions $F_l(kr)$ it follows that $f_{kl}(r) = F_l(kr)/kr$.

The ion-electron interaction is developed as

$$\frac{Z_p}{|\boldsymbol{R} - \boldsymbol{r}|} = 4\pi \, Z_p \sum_{lm} \frac{1}{2l+1} \frac{r_<^l}{r_>^{l+1}} \, Y_l^{m*}(\hat{\boldsymbol{R}}) \, Y_l^m(\hat{\boldsymbol{r}}). \tag{3.64}$$

After insertion of the (3.61,63,64) into (3.60) it follows that

$$a_k = -i\frac{Z_p}{v} \sqrt{\frac{2}{\pi}} 4\pi \frac{1}{\sqrt{4\pi}} \sum_{ll' \, mm'} \frac{i^l e^{i\sigma_l}}{2l+1} \, Y_l^m(\hat{\boldsymbol{k}}) \int_{-\infty}^{\infty} dz \, e^{iqz} \, Y_{l'}^{m'*}(\hat{\boldsymbol{R}})$$

$$\times \int_0^{4\pi} Y_l^{m*}(\hat{\boldsymbol{r}}) \, Y_{l'}^{m'}(\hat{\boldsymbol{r}}) \, d\hat{r} \int_0^{\infty} dr \, r^2 \frac{r_<^l}{r_>^{l+1}} \, f_{kl}(r) \, R_{ns}(r). \tag{3.65}$$

This expression can be simplified using the orthogonality relation of the spherical harmonics. Thus, after some algebra the transition probability $P_k = |a_k|^2$ is evaluated as

$$P_k(b) = \frac{Z_p^2}{2\pi^2 v^2} \sum_{ll' \, mm'} i^{l-l'} \, e^{i(\sigma_l - \sigma_{l'})} N_{lm} N_{l'm'}$$

$$\times \, P_l^{|m|}(\theta) \, P_{l'}^{|m'|}(\theta) \, e^{i\,(m-m')\,\varphi} \, I_{klm}^*(b) \, I_{kl'm'}(b), \tag{3.66}$$

where $N_{lm} = (l - |m|)!/(l + |m|)!$, θ is the electron observation angle, and $P_l^m(\theta)$ are associated Legendre polynomials. Furthermore, the path integral is obtained as

$$I_{klm}(b) = \int_{-\infty}^{\infty} dz \, e^{iK_m z} \, P_l^{|m|}(\theta_R) \, J_{kl}(R) \tag{3.67}$$

with the radial integral

$$J_{kl}(R) = \int_0^{\infty} dr \, r^2 \, \frac{r_<^l}{r_>^{l+1}} \, f_{kl}(r) \, R_{ns}(r)$$

$$= \int_0^R dr \, \frac{r^{l+2}}{R^{l+1}} \, f_{kl}(r) \, R_{ns}(r) + \int_R^{\infty} dr \, \frac{R^l}{r^{l-1}} \, f_{kl}(r) \, R_{ns}(r). \tag{3.68}$$

To evaluate these integrals, it is noted that the rotation angle of the internuclear axis is obtained by means of $\theta_R = \arccos(z/R)$ where the internuclear axis is given by $R = \sqrt{z^2 + b^2}$.

Equation (3.66) contains two types of cross terms, which differ in the quantum numbers l and l' or m and m'. These cross terms cancel as summations are performed over the polar and azimuthal angles, respectively. In most experiments the azimuthal angle is not observed so that a summation over φ is necessary. The corresponding integration cancels the terms that differ in the magnetic quantum numbers m and m'. Note that the individual sum terms are equal for m and -m. It is useful to express separately the diagonal terms and the cross terms:

$$P_k(b) = \frac{Z_p^2}{2\pi^2 \, v^2} \left\{ \sum_{lm} \left| N_{lm} \, P_l^{|m|}(\theta) \, I_{klm}(b) \right|^2 + 2 \sum_{l \neq l'm} \cos(\sigma_l - \sigma_{l'}) \right.$$

$$\left. \times \, N_{lm} \, N_{l'm} \, P_l^{|m|}(\theta) P_{l'}^{|m|}(\theta) \, i^{l-l'} \, I_{klm}^*(b) \, I_{kl'm}(b) \right\}, \tag{3.69}$$

where the summation over m in the cross terms is subject to the condition $|m| \leq \min(l, l')$

To obtain the related differential cross sections, the probability is integrated over the impact parameter:

$$\frac{d\sigma}{d\epsilon \, d\Omega} = k \, 2\pi \int_0^{\infty} P_k(b) \, b \, db. \tag{3.70}$$

Hence, one obtains the double-differential cross section for electron emission that can be compared with experimental results.

3.3.3 Dipole Approximation

To obtain some insight into the role of the first two terms in the partial wave expansion, we consider the dipole approximation where $l \leq 1$. From (3.69) we obtain

$$
\begin{aligned}
P_k = {} & \frac{Z_p^2}{2\pi^2\,v^2} \left(|I_{0,0}|^2 + |I_{1,0}|^2 \cos^2\theta \right) \\
& + \frac{1}{2}\,|I_{1,1}|^2 \sin^2\theta + 2\cos(\Delta\sigma) + \frac{1}{2}\,|I_{0,0}\,I_{1,0}|\cos\theta.
\end{aligned}
\tag{3.71}
$$

The last term describes the interference between the monopole and dipole term. It should be realized that this interference introduces an asymmetry with respect to $90°$ into the angular distribution of the ejected electrons.

To evaluate (3.71) we use the Legendre polynomials

$$
P_0^0 = 1, \quad P_1^0 = \cos\theta_R = z/R, \quad P_1^1 = \sin\theta_R = b/R,
\tag{3.72}
$$

where $R = \sqrt{z^2 + b^2}$ is again the internuclear distance. Hence, we obtain the integrals

$$
I_{0,0} = 2 \int_0^\infty \cos(K_m\,z)\,J_0(R)\,dz,
$$

$$
I_{1,0} = 2\,\mathrm{i} \int_0^\infty \sin(K_m\,z)\,\frac{z}{R}J_1(R)\,dz,
\tag{3.73}
$$

$$
I_{1,\pm 1} = 2 \int_0^\infty \cos(K_m\,z)\,\frac{b}{R}J_1(R)\,dz,
$$

where

$$
J_0 = \int_0^R \frac{r^2}{R}f_{k0}\,R_{ns}\,dr + \int_R^\infty r\,f_{k0}\,R_{ns}\,dr,
$$

$$
J_1 = \int_0^R \frac{r^3}{R^2}\,f_{k1}\,R_{ns}\,dr + \int_R^\infty R\,f_{k1}\,R_{ns}\,dr.
\tag{3.74}
$$

In (3.71) the interference term is determined by the phases of the continuum wave functions. For Coulomb waves these phases are given by

$$
\Delta\sigma = \arctan\left(\frac{-Z_t}{k}\right)
\tag{3.75}
$$

so that

$$
\cos(\Delta\sigma) = \left[1 + \left(\frac{Z_t}{k}\right)^2\right]^{-\frac{1}{2}},
\tag{3.76}
$$

where Z_t is the nuclear charge of the target. It is seen that the interference disappears at asymptotically small electron momenta k, whereas it maximizes at high k values. It should be kept in mind, however, that the cancellation of the monopole-dipole interference is a specific property of the hydrogenic target atom [*Briggs* and *Day* (1980)]. For multielectron targets this interference does not cancel so that the angular distribution of the electrons retain to be asymmetric at the zero-energy limit.

3.3.4 Free Electron Peaking Approximation

A simple expression for the triple differential electron emission probabilities as a function of the impact parameter has been derived by *Hansen* and *Kocbach* (1989). The derivation is based on various model assumptions including the semiclassical approximation. The outgoing electron is treated as a plane wave involving a free-electron approximation. Moreover, the model involves a peaking approximation which causes the cross section to tend to infinity as the electron energy approaches zero. In the following, this singularity will be removed by performing an empirical modification of the formula, see also [*Rudd* et al. (1996)].

In the semiclassical approximation, the projectile passes the target atom at a certain impact parameter b resulting in a scattering of the projectile into the solid angle $d\Omega_p$. Thus, the triple differential ionization cross section is given by

$$\frac{d^3\sigma}{d\Omega_p\, d\Omega\, d\epsilon} = \frac{d\sigma_R}{d\Omega_p}\, P(b, \theta, \epsilon), \tag{3.77}$$

where $d\sigma_R/d\Omega_p$ is the Rutherford cross section from (3.1) which governs the projectile-target scattering and $P(b, \theta, \epsilon)$ is the impact-parameter dependent probability for the emission of an electron into the solid angle $d\Omega$ and the energy interval $d\epsilon$.

We modified the previous evaluation of *Hansen* and *Kocbach* (1989) by inserting a few missing constants originating from the Bethe integral and the normalization of the initial hydrogenic 1s wave function $\varphi_{1s} = \pi^{1/2}\alpha^{3/2}\,e^{-\alpha r}$ and utilized the relation $dk^3 = k\, d\Omega\, d\epsilon$. Hence, we obtained the following expression, which is nearly the same as that given by *Hansen* and *Kocbach* (1989):

$$P(b, \theta, \epsilon) = \frac{Z_p^2}{v^2}\, \frac{8\alpha^5 b^2}{\pi^2 k_c^2}\, \frac{\mathcal{K}_1^2\left(b\left[\alpha^2 + (K_m - k\cos\theta)^2\right]^{1/2}\right)}{\alpha^2 + (K_m - k\cos\theta)^2}, \tag{3.78}$$

where Z_p is the projectile charge, v is the projectile velocity, $K_m = (\alpha^2 + k^2)/2v$ is the minimum momentum transfer (well known in connection with the Born approximation) and $k = (2\epsilon)^{1/2}$ is the momentum of the outgoing electron. The momentum k_c will be defined later. The impact-parameter dependence is governed by the modified Bessel function $\mathcal{K}_1$. The parameter

α, which is a measure of the mean velocity of the bound electron, is obtained from the well-known relation $\alpha = (2E_b)^{1/2}$ for hydrogen-like ions, where E_b is the corresponding ionization energy.

To obtain double differential cross sections, (3.78) is integrated over the impact parameter using the well-known expression (3.70). This integration can be performed analytically, yielding an expression of remarkable simplicity:

$$\frac{d\sigma}{d\Omega\,d\epsilon} = \frac{Z_p^2}{v^2}\,\frac{32}{3\pi\,\alpha\,k_c^3}\left(\frac{1}{1 + \left(\tilde{K}_m - \tilde{k}\cos\theta\right)^2}\right)^3, \qquad (3.79)$$

where $\tilde{K}_m = K_m/\alpha$ and $\tilde{k} = k/\alpha$ are normalized momenta. The function in large parenthesis describes the binary encounter maximum that resembles a Lorentzian whose width is governed by α. This function determines the dependence on the electron emission angle.

The expression in (3.79) can be further integrated over the electron emission angle to obtain the ionization cross section differential in energy only:

$$\frac{d\sigma}{d\epsilon} = \frac{Z_p^2}{v^2}\,\frac{16}{3\,\alpha\,k_c^3\,\tilde{k}}\left[\arctan\left(\frac{2\tilde{k}}{1 + \tilde{K}_m^2 - \tilde{k}^2}\right)\right.$$

$$+\,\frac{5\left(\tilde{K}_m + \tilde{k}\right) + 3\left(\tilde{K}_m + \tilde{k}\right)^3}{2\left[1 + \left(\tilde{K}_m + \tilde{k}\right)^2\right]^2}$$

$$\left.-\,\frac{5\left(\tilde{K}_m - \tilde{k}\right) + 3\left(\tilde{K}_m - \tilde{k}\right)^3}{2\left[1 + \left(\tilde{K}_m - \tilde{k}\right)^2\right]^2}\right] \qquad (3.80)$$

Hence, a relatively simple analytic expression is obtained for the single-differential cross section for electron emission.

The original equations by *Hansen* and *Kocbach* (1989) imply $k_c = k$. However, the present expressions provide a certain flexibility as the parameter k_c may be adjusted to values different from that of k. We note that $k_c = k$ follows from the peaking approximation which neglects the momenta of the bound electrons in comparison to the momentum for the outgoing electron. Hence, the peaking approximation produces a singularity in the low-energy limit for the electron.

We removed this singularity by arbitrarily setting $k_c^2 = k^2 + p_o^2$ where p_o is a quantity governed by the mean momentum of the bound electron. We have chosen a semiempirical expression for k_c to fit the model results to those of the Born approximation [*Landau* and *Lifschitz* (1958)]:

$$k_c = \left[k^2 + \alpha^2 \frac{3}{2} \left(\ln \frac{2v^2}{\alpha^2}\right)^{-2/3}\right]^{1/2}. \tag{3.81}$$

This expression for k_c is consistent with that given previously in (3.6). It is seen that the mean momentum p_o is approximately equal to α which represents the mean momentum of the bound electron. The term including the logarithm follows from the Born approximation describing soft collisions for which dipole transitions are important [*Bethe* (1930)].

Results for double and single-differential cross sections obtained from the free-electron peaking approximation (FEPA) are shown in Fig. 2.4b. The data compare well with corresponding cross sections in Fig. 2.4a evaluated by means of the first Born approximation [*Landau* and *Lifschitz* (1958), *Kuyatt* and *Jorgensen* (1963)].

The double-differential cross section exhibits prominent binary-encounter peaks. It is expected that the model yields accurate results for the region where the binary-encounter maximum is dominant. Accordingly, good agreement is obtained for the single-differential cross sections that are primarily determined by the binary-encounter mechanism. However, problems are expected in the description of the soft collisions. Indeed, the results from the FEPA and B1 differ at low electron energies (Fig. 2.4). Further examples for calculations by the free-electron peaking approximation will be presented in forthcoming sections.

3.4 Born Approximations

In the framework of quantum mechanics, the first-order of the Born series (B1) describes the heavy particle as a plane wave in the entrance and exit channels. The electron is described by bound and continuum wave functions of the target in the entrance and exit channels, respectively. These wave functions are characteristic for the first term of a Born series where the electron is transferred to a continuum state of the target.

A further approximation of B1 can be obtained if the electron in the final state is described by a simple plane wave, assuming that the three particles in the exit channel are free. This Born approximation with a final plane wave (B1-PW), describing the zero-center case shall be treated first.

3.4.1 Free-Electron Approximation

Denote the initial stationary wave function as φ_i which describes the electron bound to the target nucleus. The initial and final wave functions are

$$\Phi_i^B = \frac{1}{(2\pi)^{3/2}} \exp[i\boldsymbol{K}_i \cdot \boldsymbol{R}_i]\, \varphi_i(\boldsymbol{x}) \tag{3.82}$$

and

$$\Phi_f^{PW} = \frac{1}{(2\pi)^3} \, \exp[\mathrm{i}\, \boldsymbol{K}_f \cdot \boldsymbol{R}_i] \, \exp[\mathrm{i}\, \boldsymbol{k} \cdot \boldsymbol{x}], \tag{3.83}$$

respectively. In (3.82,83), $\boldsymbol{R}_i$ is the position of the projectile measured from the center of mass of the target, $\boldsymbol{K}_i(\boldsymbol{K}_f)$ is the corresponding initial (final) wavevector of the incident (scattered) reduced particle, $\boldsymbol{x}$ is the electron-target nucleus position vector, and $\boldsymbol{k}$ is the momentum of the reduced particle of mass $\mu_T = M_T/(M_T + 1)$ seen from the center of mass of the target, with M_T the mass of the target nucleus.

Hereafter, in any of the approximations used, we will not take into account the internuclear potential. It has been shown that this interaction can be treated exactly within the impact parameter approximation. The only effect on the transition amplitude is the production of a phase factor, depending on the impact parameter, which does not contribute to the cross section [*Belkić et al.* (1979), *Fainstein et al.* (1988a)].

It is easy to show that the scattering matrix element T_{if}^{B1-PW} can be written as

$$\begin{aligned} T_{if}^{B1-PW}(\boldsymbol{K}) &= \langle \Phi_f^{PW} | - \tfrac{Z_p}{s} | \Phi_i^B \rangle \\ &= \frac{1}{(2\pi)^{3/2}} \, \tilde{V}_P(\boldsymbol{K}) \, \tilde{\varphi}_i(\boldsymbol{k} - \mu_T \boldsymbol{K}) \end{aligned} \tag{3.84}$$

where Z_p is the projectile nuclear charge, s is the electron-projectile nucleus position vector and $\boldsymbol{K} = \boldsymbol{K}_i - \boldsymbol{K}_f$ is the momentum transfer in the center of mass of the system. It is clear that as $M_T \gg 1$, $\mu_T \cong 1$ and $\boldsymbol{k}$ is the momentum of the electron relative to the target nucleus. In (3.84) $\tilde{\varphi}_i(\boldsymbol{K}) = (2\pi)^{-3/2} \int d^3r \, \exp(-\mathrm{i}\,\boldsymbol{K}\cdot\boldsymbol{r}) \, f(\boldsymbol{r})$ indicates the Fourier transform of φ_i and $\tilde{V}_P(\boldsymbol{K}) = -2^{1/2}\pi^{-1/2} Z_p/K^2$ is the Fourier transform of the perturbing potential $V_P(\boldsymbol{s}) = -Z_p/s$.

Using the equations for momentum conservation in the total system, it is easy to see that $\boldsymbol{k}_i = \boldsymbol{k} - \mu_T \boldsymbol{K} \cong \boldsymbol{k} - \boldsymbol{K}$ stands for the momentum of the electron in its initial state. Hence, $\tilde{\varphi}_i(\boldsymbol{k}_i)$ describes the initial electron momentum distribution. Equation (3.84) can thus be interpreted in the following way: The electron in an initial distribution $\tilde{\varphi}_i(\boldsymbol{k}_i)$ receives a momentum $\boldsymbol{K}$ due to the collision with the projectile. This transfer is determined by the quantity $\tilde{\varphi}_i(\boldsymbol{k} - \boldsymbol{K})$, weighted by the transform of the perturbing potential $\tilde{V}(\boldsymbol{K})$. As a result, the electron is ejected with a final momentum $\boldsymbol{k} = \boldsymbol{k}_i + \boldsymbol{K}$.

Taking into account only projectiles that are scattered into a forward cone [$\boldsymbol{K}_i \approx \boldsymbol{K}_f$; *Belkić* (1978)]:

$$\boldsymbol{K} = \boldsymbol{\eta} + \frac{\Delta\epsilon}{v^2}\boldsymbol{v}, \quad \Delta\epsilon = \frac{k^2}{2} - \epsilon_i, \tag{3.85}$$

where ϵ_i is the initial orbital energy of the electron. In 3.85), $\boldsymbol{\eta}$ is the transverse momentum transfer and $\boldsymbol{v}$ is defined by $\boldsymbol{v} = \boldsymbol{K}_i/\nu_i$, with $\nu_i = M_P(M_T + 1)/(M_T + M_P + 1)$ being the reduced mass of the incident particle and M_P the mass of the projectile nucleus.

If we consider an initial $1s$ state,

$$\tilde{\varphi}_i(|\boldsymbol{k} - \boldsymbol{K}|) = \tilde{\varphi}_i(k_i) = \frac{2^{3/2} Z_t^{5/2}}{\pi \left(Z_t^2 + k_i^2\right)^2} \tag{3.86}$$

with Z_t the target nuclear charge. The momentum distribution thus shows a peak at $k_i = 0$ ($\boldsymbol{k} - \boldsymbol{K} \cong 0$), which, in order $1/M_T$ corresponds to the values of $\boldsymbol{k}$ situated in the binary sphere:

$$\boldsymbol{k} \cdot \hat{\boldsymbol{v}} - \frac{\Delta \epsilon}{v} = 0. \tag{3.87}$$

With (3.85) this expression can approximately be written in the more familiar form

$$\epsilon_{BE} = \frac{k^2}{2} \cong 4T \cos^2 \theta - 2 |\epsilon_i|, \tag{3.88}$$

where ϵ_{BE} is the electron energy in the BE peak. Equation (3.88) is identical with (2.1) except that twice the absolute value of the binding energy is subtracted.

The double-differential cross section is obtained as

$$\frac{d\sigma}{d\epsilon\, d\Omega} = (2\pi)^4\, v_i^2\, k \int_{4\pi} d\Omega_K \; |T_{if}(\boldsymbol{K})|^2 = (2\pi)^4\, \frac{k}{v^2} \int d^2\boldsymbol{\eta}\; |T_{if}(\boldsymbol{K})|^2 \, . \tag{3.89}$$

The solid angle Ω_K is subtended by the momentum transfer $\boldsymbol{K}$.

The double-differential cross section $d\sigma/d\epsilon d\Omega$ reveals structures of the Compton profile [*Bell* et al. (1983), *Fainstein* et al. (1989a)]. Inserting (3.84) into (3.89) one obtains

$$\frac{d\sigma^{B1-PW}}{d\epsilon d\Omega} = 4Z_p^2 \frac{k}{v^2} \int d^2\boldsymbol{\eta} \frac{|\varphi(\boldsymbol{k}_i)|^2}{K^4}. \tag{3.90}$$

For further evaluation we substitute the integration variable $\boldsymbol{\eta}$ by $k_{i\perp}$ where $k_{i\perp}$ is the component of $\boldsymbol{k}_i$ perpendicular to the direction given by $\boldsymbol{K}_i$. Also, as was pointed out in Sect. 3.1.2, the initial momentum distribution $\tilde{\varphi}_i(\boldsymbol{k}_i)$ is related to the Compton profile $J(k_{iz}) = \int d^2 k_{i\perp} \; |\tilde{\varphi}_i(\boldsymbol{k}_i)|^2$. Recalling that $\boldsymbol{K} = \boldsymbol{k} - \boldsymbol{k}_i$ and using the fact that $\tilde{\varphi}_i(\boldsymbol{k}_i)$ maximizes at $\boldsymbol{k}_i = 0$, we perform a peaking approximation by setting $\boldsymbol{K} = \boldsymbol{k}$. Hence, the double differential cross section corresponding to the B1-PW model can be approximated by the simple formula [*Bell* et al. (1983)]:

$$\frac{d\sigma^{B1-PW}}{d\epsilon d\Omega} = \frac{4Z_p^2}{v^2 k^3} \, J(k_{iz}). \tag{3.91}$$

This expression describes the binary-encounter peak quite well. It can readily be verified that the location of the binary-encounter peak is determined by the condition that the Compton profile maximizes at $k_{iz} = 0$. It is found to be shifted to lower electron velocities due to the additional term containing $2|\epsilon_i|$ in (3.88).

Using the Compton profile of a hydrogenic $1s$ wave function, taking $Z_t = \alpha$ and considering the relationship $k_{iz} = k \cos \theta - K_m$ (which can be derived from (3.85) and where $K_m = \Delta \epsilon / v$) in (3.89), we recover (3.79) in Sect. 3.3.4 where the semiclassical approximation is treated. This shows that within the framework of energetic ion-atom collisions the B1 approximation and the semiclassical approximation are closely related. Accordingly, as the semiclassical model in Sect. 3.3.4 we refer to the present model, yielding (3.91), to as the free-electron peaking approximation.

3.4.2 First-Order Born Approximation

For soft collisions ($K \ll K_i$), the electrons will preferably be ejected with small momentum k. Thus, the electrons will feel the potential of the target nucleus in the exit channel and its simple description by a plane wave should be improved. To adequately describe the electron traveling in a continuum state of the target it is necessary to modify the final wave function in the exit channel:

$$\Phi_f^B = \frac{\exp\left(\mathrm{i} \boldsymbol{K}_f \cdot \boldsymbol{R}_i\right)}{(2\pi)^{3/2}} \, \varphi_f^-$$

(3.92)

with

$$\varphi_f^- = \frac{\exp\left(\mathrm{i} \boldsymbol{k} \cdot \boldsymbol{x}\right)}{(2\pi)^{3/2}} \, N(\xi) \, {}_1F_1[-\mathrm{i}\xi; 1; -\mathrm{i}(kx + \boldsymbol{k} \cdot \boldsymbol{x})],$$

(3.93)

where $\xi = Z_t/k$, and $N(a) = \exp(a\pi/2)\Gamma(1 + \mathrm{i}a)$ is the normalization factor of the continuum wave function, which has been chosen to satisfy incoming boundary conditions. The use of the wave functions given by (3.82,92) in the calculation of the T-matrix element yields the B1 approximation. We obtain thus,

$$T_{if}^{B1}(\boldsymbol{K}) = \frac{1}{(2\pi)^{3/2}} \tilde{V}_P(\boldsymbol{K}) \int d^3x \ \exp\left(\mathrm{i} \boldsymbol{K} \cdot \boldsymbol{x}\right) \left(\varphi_f^-(\boldsymbol{x})\right)^* \varphi_i(\boldsymbol{x})$$

$$= \frac{1}{(2\pi)^{3/2}} \tilde{V}_P(\boldsymbol{K}) \, F_{0k}(-\boldsymbol{K}),$$

(3.94)

where $F_{0k}(-\boldsymbol{K})$ is the form factor. It can readily be shown from (3.94) that in the limit $\xi \to 0$ we recover the result given by (3.84), corresponding to the B1-PW approximation.

It is useful to define the amplitude $R_{if}(\boldsymbol{\eta})$ as a function of the transverse momentum transfer, using the relationship

$$R_{if}(\boldsymbol{\eta}) = -\mathrm{i} \frac{4\pi^2}{v} T_{if}(\boldsymbol{K}).$$

(3.95)

Combining (3.89,95) we obtain

$$\frac{d\sigma}{d\epsilon \, d\Omega} = k \int d^2\eta \ |R_{if}(\boldsymbol{\eta})|^2,$$

(3.96)

showing that $|R_{if}(\eta)|^2$ is a useful quantity for the calculation of the double-differential cross section.

When we consider ionization from a $1s$ ground state in the B1 approximation, we obtain [*Belkić* (1978), *Crothers* and *McCann* (1983)]:

$$|R_{if}^{B1}(\eta)|^2 = 4|N(\xi)|^2 \exp\left[-2\xi \tan^{-1}\left(\frac{2Z_t k}{Z_t^2 + K^2 + k^2}\right)\right]$$

$$\times \frac{[(K^2 - \boldsymbol{K} \cdot \boldsymbol{k})^2 + (\boldsymbol{K} \cdot \boldsymbol{k})^2 \xi^2]}{[Z_t^2 + (K+k)^2][Z_t^2 + (K-k)^2]} \; |R_{if}^{B1-PW}(\eta)|^2, \tag{3.97}$$

where $R_{if}^{B1-PW}(\eta)$ is the amplitude corresponding to the B1-PW approximation. It is given by the expression

$$R_{if}^{B1-PW}(\eta) = i\frac{2^{5/2}}{\pi} \frac{Z_p Z_t^{5/2}}{v K^2 [Z_t^2 + (\boldsymbol{k} - \boldsymbol{K})^2]^2}. \tag{3.98}$$

It is recalled that the B1 approximation applies to the single-center case involving a target-centered continuum wave function. In particular, this approximation is well-suited for describing the dipole-type transitions producing the soft-collision peak and the two-body interactions creating the binary-encounter maximum by bare projectiles.

In the past, scaled hydrogenic wave functions have often been used in conjunction with the B1 approximation. This approximation has the advantage that the theoretical results obtained for a given target species can be scaled to other target atoms [*Rudd* et al. (1966a,b)]. Within the framework of the first-Born approximation an analytic expression for the ionization of a hydrogen-like $1s$ electron has been evaluated by several authors [*Massey* and *Mohr* (1933), *Mott* and *Massey* (1949), *Massey* (1956), *Landau* and *Lifschitz* (1958)]. Some results are to be taken with care as they include misprints. The correct result from *Landau* and *Lifschitz* (1958) shall be given here as it is used several times to obtain theoretical results for electron emission by ion impact.

The cross section is differential in the momentum transfer K by the projectile and the momentum k and solid angle Ω of the emitted electron [*Landau* and *Lifschitz* (1958), *Kuyatt* and *Jorgensen* (1963)]. This triple-differential cross section is expressed as

$$\frac{d\sigma}{dK\,dk\,d\Omega} = \frac{2^8\,k}{K\,v^2} \frac{e^{-\frac{2}{k}\tan^{-1}[2k/(K^2 - k^2 + 1)]}}{\left[(K+k)^2 + 1\right]\left[(K-k)^2 + 1\right]\left(1 - e^{-\frac{2\pi}{k}}\right)}$$

$$\times \frac{CD^3 + 4CDE^2 - 4BD^2E - BE^3 + 2AD^3 + 3ADE^2}{(D^2 - E^2)^{7/2}}, \tag{3.99}$$

where

$$A = K^2 - 2K_m\,k\cos\theta + (k^2 + 1)\left(\frac{K_m}{K}\right)^2\cos^2\theta,$$

$$B = 2\left(K^2 - K_m^2\right)^{1/2} k \sin\theta - \left(k^2 + 1\right)\frac{2K_m}{K^2}\left(K^2 - K_m^2\right)^{1/2}\sin\theta\cos\theta,$$

$$C = \left(k^2 + 1\right)\frac{K^2 - K_m^2}{K^2}\sin^2\theta,$$

$$D = K^2 - 2K_m\, k\cos\theta + k^2 + 1,$$

$$E = 2k\left(K^2 - K_m^2\right)^{1/2}\sin\theta.$$

As usual, v is the projectile velocity and K_m is the minimum momentum transfer.

To obtain double differential cross sections relevant for the results discussed in this work the above expression has to be integrated with respect to the momentum transfer, Sect. 3.7.1. The resulting double-differential cross section $d\sigma/dk\,d\Omega$ involves a useful scaling property given by *Rudd* et al. (1966a):

$$\frac{d\sigma}{dk\,d\Omega}\left(\epsilon, E_p, E_b\right) = n\left(\frac{U_b}{E_b}\right)^3 \frac{d\sigma}{dk\,d\Omega}\left(\epsilon U_b/E_b,\, E_p U_b/E_b,\, U_b\right), \quad (3.100)$$

where n is the number of electrons with the ionization potential E_b. The corresponding ionization potential of the hydrogenic 1s electron is $U_b = 0.5$ a.u. As before, ϵ and E_p are the energies of the electron and projectile, respectively. It is seen that both ϵ and E_p are affected by the dimensionless scaling factor U_b/E_b.

However, the hydrogenic scaling procedure strongly underestimates the cross sections for electron emission at backward angles. This is because the backscattering mechanism indicated in Fig. 2.3d is not properly described. The underestimation of the cross sections following from the use of hydrogenic wave functions will be discussed in more detail further on in the experimental sections.

Consequently, at backward angles, the agreement between theory and experiment can significantly be improved when realistic potentials are used to describe the screened target center. *Madison* (1973) and *Manson* et al. (1975) have shown that good agreement between experiment and theory can also be achieved at backward angles when Hartree-Slater potentials are utilized to evaluate the initial bound and final continuum wave functions. However, it should be emphasized that this agreement obtained for proton impact is not necessarily reproduced for highly charged ions.

In the following, the theoretical method introduced by *Madison* (1973) is treated in more detail. The main idea is to describe the target atom by an independent electron model, so that each electron moves in the field of a single center potential.

The active electron-continuum wave function is described by its partial wave expansion

$$\varphi_f^-(\boldsymbol{x}) = \sum_{l,m} i^l \, e^{-i\delta_l} \frac{u_{\epsilon l}(x)}{x\sqrt{k}} \, Y_l^m(\hat{\boldsymbol{k}}) \, [Y_l^m(\hat{\boldsymbol{x}})]^* \tag{3.101}$$

which is normalized to the energy ϵ through the condition

$$\lim_{x\to\infty} u_{\epsilon l}(x) = \sqrt{\frac{2}{\pi k}} \, \sin\left[kx + \frac{1}{k}\ln(2kx) - \frac{l\pi}{2} + \delta_l \right]. \tag{3.102}$$

Let us write the initial bound wave function as

$$\varphi_i(\boldsymbol{x}) = \frac{u_{n_i l_i}(x)}{x} \, Y_{l_i}^{m_i}(\hat{\boldsymbol{x}}) \tag{3.103}$$

and exp[i $\boldsymbol{K} \cdot \boldsymbol{x}$] by its partial wave expansion

$$\exp\left[i\boldsymbol{K} \cdot \boldsymbol{x} \right] = 4\pi \sum_{l,m} i^l j_l(Kx) \, Y_l^m(\hat{\boldsymbol{K}}) \, [Y_l^m(\hat{\boldsymbol{x}})]^* \,, \tag{3.104}$$

where j_l is a spherical Bessel function. Then, inserting the expressions (3.101,103,104) in (3.94) and using the definition given by (3.95), we obtain

$$R_{if}^{B1}(\boldsymbol{\eta}) = i\frac{4\pi^{1/2} Z_p}{v \, K^2 \sqrt{k}} \sum_{lm} \sum_{l'm'} (-1)^{l+m'} i^{l+l'} \, e^{i\delta_l}$$

$$\times \left[(2l'+1)(2l+1)(2l_i+1) \right]^{1/2} \begin{pmatrix} l' & l & l_i \\ 0 & 0 & 0 \end{pmatrix} \tag{3.105}$$

$$\times \begin{pmatrix} l' & l & l_i \\ -m' & m & m_i \end{pmatrix} \left[Y_l^m(\hat{\boldsymbol{k}}) \right]^* \left[Y_{l'}^{m'}(\hat{\boldsymbol{K}}) \right] f_{l,l',l_i}^{\epsilon,n_i}(K)$$

with

$$f_{l,l',l_i}^{\epsilon,n_i}(K) = \int_0^\infty dx \, u_{\epsilon l}^*(x) \, j_{l'}(Kx) \, u_{n_i l_i}(x) \tag{3.106}$$

and $\begin{pmatrix} a & b & c \\ d & e & f \end{pmatrix}$ being the Wigner $3 - j$ symbol.

In (3.106) the one-electron orbitals $u_{nl}(x)$ are usually taken from the Hartree-Fock-Slater form tabulated by *Herman* and *Skillman* (1963). They are solutions of the radial Schrödinger equation with the central potential $V_T(x)$,

$$\left(\frac{1}{2}\frac{d^2}{dx^2} - V_T(x) - \frac{l(l+1)}{2x^2} + \epsilon_{nl} \right) u_{nl}(x) = 0, \tag{3.107}$$

where ϵ_{nl} is the corresponding electron energy. The continuum orbitals $u_{\epsilon l}$ are derived from the same radial Schrödinger equation (3.107) as used for the bound states. *Salin* (1989) applied an analytical approximation of the Herman-Skillman potential to derive bound and continuum states of He targets.

Averaging over initial and summing over final degenerate magnetic sub-states, double differential cross sections can be obtained using (3.95,105) in (3.89) [*Manson* et al. (1975)]:

$$\frac{d\sigma}{d\epsilon d\Omega} = \left(\frac{2Z_p}{v}\right)^2 N_{n_i l_i} \sum_{l,l',\lambda,\lambda',L} i^{l+l'+\lambda+\lambda'} e^{i(\delta_l - \delta_{l'})}$$

$$\times (2l+1)(2l'+1)(2\lambda+1)(2\lambda'+1)(2L+1) \begin{pmatrix} l' & \lambda' & l_i \\ 0 & 0 & 0 \end{pmatrix}$$

$$\times \begin{pmatrix} l & \lambda & l_i \\ 0 & 0 & 0 \end{pmatrix} \begin{pmatrix} l & l' & L \\ 0 & 0 & 0 \end{pmatrix} \begin{pmatrix} \lambda & \lambda' & L \\ 0 & 0 & 0 \end{pmatrix} \begin{Bmatrix} \lambda & \lambda' & L \\ l' & l & l_i \end{Bmatrix}$$

$$\times \left(\int_{K_{min}}^{K_{max}} dK \frac{1}{K^3} f_{l,\lambda,l_i}^{\epsilon,n_i}(K) f_{l',\lambda,l_i}^{\epsilon,n_i}(K) P_L(\cos\theta_K) \right) P_L(\cos\theta) \quad (3.108)$$

where $N_{n_i l_i}$ is the number of electrons initially in the subshell $n_i l_i$ and $P_L(\cos\theta)$ is a Legendre Polynomial. Moreover,

$$\cos\theta_K = \frac{K^2 + K_i^2 - K_f^2}{2K_i K_f} \quad (3.109)$$

and $\begin{Bmatrix} a & b & c \\ d & e & f \end{Bmatrix}$ is the Wigner $6-j$ symbol.

The B1 approximation generally fails for electron ejection in regions other than those of the soft-collision maximum and the binary encounter peak, even when realistic continuum wave functions are utilized. For instance, at forward angles, the theory significantly underestimates the experimental data [*Manson* et al. (1975)]. This discrepancy is produced by two-center effects treated in a later section.

3.4.3 Oppenheimer-Brinkman-Kramers Approximation

In order to describe the enhancement of the experimental data at forward angles, *Dettmann* et al. (1974) interpreted the reaction using a theory of electron capture to the continuum of the projectile. Similar to charge exchange to a bound state, a second-order Born approximation is necessary to calculate the cross sections at high collision energies. The corresponding Born series is named Oppenheimer-Brinkman-Kramers series in association with the case of electron capture to a bound state [*Oppenheimer* (1928), *Brinkman* and *Kramers* (1930)], and the first- and second-order approximations are denoted by OBK1 and OBK2, respectively.

The initial wave function is taken as in the B1 model and the final wave function is chosen as

$$\Phi_f^{OBK} = \frac{1}{(2\pi)^3} \exp\left(\mathrm{i}\boldsymbol{K}_f \cdot \boldsymbol{R}_f\right) \exp\left(\mathrm{i}\boldsymbol{p} \cdot \boldsymbol{s}\right) N\left(\zeta\right) {}_1F_1[-\mathrm{i}\zeta; 1; -\mathrm{i}(ps + \boldsymbol{p} \cdot \boldsymbol{s})],$$

(3.110)

where $\zeta = Z_p/p$, $\boldsymbol{p} = \boldsymbol{k} - \boldsymbol{v}$ is the momentum of the electron with respect to the projectile, and $\boldsymbol{R}_f$ is the position of the center of mass of the projectile-electron sub-system with respect to the target nucleus. Then, the matrix elements T_{if}^{OBK1} and T_{if}^{OBK2} corresponding to the first- and second-orders of the OBK series are given by

$$T_{if}^{OBK1}(\boldsymbol{K}) = \left\langle \Phi_f^{OBK} \left| -\frac{Z_p}{s} \right| \Phi_i^B \right\rangle$$

(3.111)

and

$$T_{if}^{OBK2}(\boldsymbol{K}) = T_{if}^{OBK1}(\boldsymbol{K}) + \left\langle \Phi_f^{OBK} \left| \left(-\frac{Z_t}{x}\right) G_0^+ \left(-\frac{Z_p}{s}\right) \right| \Phi_i^{PW} \right\rangle$$

(3.112)

respectively, where G_0^+ is a free-particle Green operator. It is easy to show that

$$\left| T_{if}^{OBK1}(\boldsymbol{K}) \right|^2 = |N(\zeta)|^2 \left| T_{if}^{B1-PW}(\boldsymbol{K}) \right|^2,$$

(3.113)

where

$$|N(\zeta)|^2 = 2\pi\zeta \left(1 - e^{-2\pi\zeta}\right)^{-1}$$

(3.114)

so that $|N(\zeta)|^2 = (2\pi Z_p)/p$ as $p \to 0$. Hence, a cusp-shaped peak appears in the double-differential cross sections for electrons ejected with a velocity close to the projectile velocity. This structure is known as the electron capture to the continuum (ECC) peak.

It should be noted that $|N(\zeta)|^2$ gives the density of continuum states around the projectile. The quantity $|N(\zeta)|^2$ appears also in the second term of the right-hand side of (3.112) giving a strong influence on the ECC peak. *Dettmann* et al. (1974) have shown that for high impact velocities and $p \cong 0$, the double-differential cross section based on T_{if}^{OBK2} is reduced by a factor of 0.3 in comparison with the double-differential cross section corresponding to T_{if}^{OBK1}. The fact that the second-order term of T_{if}^{OBK2} affects the ECC peak does not mean that the quantum mechanical reaction associated with the classical two-step intermediate mechanism [*Thomas* (1927)] will play a role. *Dettmann* et al. (1974) estimated the contribution of this mechanism on the ECC peak, showing that it is negligible except at asymptotic high-collision velocities.

Higher orders of the OBK series could also give important contributions at intermediate and high-collision velocities. At least, this is the case for electron capture to bound states [*Miraglia* et al. (1980)].

The ionization reaction was also studied using the Fadeev's formalism, [*Fadeev* (1960)]. The first term of the corresponding Neumann expansion was calculated by *Macek* (1970), showing that the final wave function can be written as

$$\Phi_f^F = \Phi_f^B + \Phi_f^{OBK} - \Phi_f^{PW} \tag{3.115}$$

so that the scattering matrix element results

$$T_{if}^F = T_{if}^{B1} + T_{if}^{OBK1} - T_{if}^{B1-PW}. \tag{3.116}$$

However, the corresponding double-differential cross sections for 2-MeV $He^{2+} + He$ collisions show interference structures at forward angles, which were not observed experimentally [*Duncan* et al. (1977)]. These interferences arise from the fact that the reaction is described as a coherent sum of different mechanisms, such as excitation to the continuum of the target and capture to the continuum of the projectile.

3.5 Distorted Wave Models

The pioneering work by *Salin* (1969, 1972) introduced for the first time a rigorous theoretical description of the two-center phenomena. Salin extended the theory by *Rudge* and *Seaton* (1964), previously formulated to study ionization in electron-atom collisions, to the case of proton impact. In order to describe the ejected electron moving in the presence of both the target and projectile fields, the final wave function $\varphi_{\bar{f}}$ was distorted by a multiplicative projectile continuum factor,

$$\mathcal{F}(s) = N\left(\frac{Z}{p}\right) \, {}_1F_1\left(-\frac{iZ}{p}; 1; -i\left(ps + \boldsymbol{p} \cdot \boldsymbol{s}\right)\right), \tag{3.117}$$

where $\boldsymbol{p} = \boldsymbol{k} - \boldsymbol{v}$ is the final electron velocity measured from a reference frame fixed at the projectile, and Z is a dynamically effective charge, defined as

$$Z = 1 - \frac{p}{v}. \tag{3.118}$$

When the electron moves with a velocity close to the projectile ($p \cong 0$), it travels in a continuum state of the proton which lies just above the threshold of the corresponding continuum spectrum. We should recall that the electron also feels the field of the target nucleus. For electrons ejected in soft collisions ($k \ll v$), $Z \cong 0$, and (3.117) indicates that they move only in the field of the target nucleus. A similar behavior is produced in the binary-encounter region, for which $p = |\boldsymbol{k} - \boldsymbol{v}| \cong v$ and so $Z \cong 0$. Perhaps, the most important conclusion obtained by *Salin* (1969) is that the resulting amplitude satisfies the equation

$$|R_{if}^s(\boldsymbol{\eta})|^2 = \left|N\left(\frac{Z}{p}\right)\right|^2 |R_{if}^{B1}(\boldsymbol{\eta})|^2. \tag{3.119}$$

The quantity

$$|N(Z/p)|^2 = \frac{2\pi Z}{p} \left(1 - e^{-2\pi Z/p}\right)^{-1}, \qquad (3.120)$$

known as the Salin factor, shows a divergence of the form $|N(Z/p)|^2 \to 2\pi/p$ as $p \to 0$. Thus, under these conditions, the double-differential cross section $d\sigma/d\epsilon\, d\Omega$ presents the electron capture to the continuum (ECC) peak. It is obvious that the condition $\boldsymbol{k} \cong \boldsymbol{v}$ is produced for electrons ejected with small scattering angle. The Salin theory thus gives a theoretical explanation of the hump observed by *Rudd* et al. (1966a) when double-differential cross sections were measured at an electron scattering angle of $\theta = 10°$ for impact of 300-keV protons on H_2.

Madison (1973) has included the Salin factor in his Hartree-Fock B1 calculations to describe the ionization of He targets by impact of 100-keV and 200-keV protons. This multiplicative factor improves the agreement between theoretical and experimental double-differential cross sections at small electron angles. However, the agreement at large angles becomes less satisfactory.

3.5.1 Continuum Distorted Wave Approximation

The B1 approximation treats the projectile potential $(-Z_p/s)$ like a perturbation and will not satisfactorily describe the ionization reaction if s is small and/or Z_p is large. The failure is noticeable for electrons ejected in the forward direction with $\boldsymbol{k} \approx \boldsymbol{v}$. Thus, we need to include this Coulomb potential in the initial and final wave functions in order to obtain a perturbing residual potential. With this goal in mind the Continuum Distorted Wave (CDW) approximation introduced by *Cheshire* (1964) to describe electron capture was extended to the ionization reaction [*Belkić* (1978)]. The CDW approximation makes use of the fact that the electron can be described exactly when the aggregates of the collision system are nearly separated [*Salin* (1969), *Belkić* (1978)]. This asymptotic behavior corresponds to

$$\lim_{R\cdot\hat{v}\to-\infty} \Psi_i^+ \approx \Phi_i^B \exp\left[-i\frac{Z_p}{v} \ln(vs + \boldsymbol{v}\cdot\boldsymbol{s})\right] \qquad (3.121)$$

in the entrance channel, and

$$\lim_{R\cdot\hat{v},x,s\to+\infty} \Psi_f^- \approx (2\pi)^{-3} \exp\left(i\,\boldsymbol{K}_f\cdot\boldsymbol{R}_i + i\,\boldsymbol{k}\cdot\boldsymbol{x}\right)$$

$$\times \exp\left[i\frac{Z_t}{k} \ln(kx + \boldsymbol{k}\cdot\boldsymbol{x}) + i\frac{Z_p}{p} \ln(ps + \boldsymbol{p}\cdot\boldsymbol{s})\right] \qquad (3.122)$$

in the exit channel. In (3.121,122), $\Psi_i^+ \left(\Psi_f^-\right)$ is the exact wave function with correct outgoing (incoming) boundary conditions, and $\boldsymbol{R}$ is the internuclear vector.

Belkić (1978) proposed initial and final distorted wave functions given by the expressions

$$\chi_i^{CDW,+} = \Phi_i^B \, L_i^+(s) = \Phi_i^B \, N^*(\nu) \, _1F_1\left(i\nu; 1; i(vs + \boldsymbol{v} \cdot \boldsymbol{s})\right) \qquad (3.123)$$

and

$$\chi_f^{CDW,-} = \Phi_f^B \, L_f^-(s) = \Phi_f^B \, N(\zeta) \, _1F_1\left(-i\zeta; 1; -i(ps + \boldsymbol{p} \cdot \boldsymbol{s})\right), (3.124)$$

respectively, where we have defined $\nu = Z_p/\mathrm{v}$ and $\zeta = Z_p/p$, as above.

It is easy to show that $\chi_i^{CDW,+}$ and $\chi_f^{CDW,-}$ satisfy the correct boundary conditions given by (3.121,122), respectively. Moreover, the potential $(-Z_p/s)$ is partially included in the description of the electron in the entrance and exit channels. In the entrance channel the electron is thus considered to move with velocity $(-\boldsymbol{v})$ with respect to the projectile frame in a continuum state of the projectile, being bound at the same time to the target nucleus. In the exit channel the electron travels in a continuum state of the combined fields of the target nucleus and projectile. The final electron continuum state is thus approximated by a function that implies the influence of the target nucleus and projectile on an equal footing.

If H and E denote the Hamiltonian and energy of the collision system, respectively, the perturbing potentials W_i and W_f result from the equations:

$$(H - E)\chi_i^{CDW,+} = W_i\chi_i^{CDW,+} = \Phi_i^B \left[\nabla_x \ln \varphi_i(\boldsymbol{x}) \cdot \nabla_s L_i^+(s)\right] \qquad (3.125)$$

and

$$(H - E)\chi_f^{CDW,-} = W_f\chi_f^{CDW,-} \qquad (3.126)$$
$$= \Phi_f^B \left[\nabla_x \ln \, _1F_1(-i\xi; 1; -i(kx + \boldsymbol{k} \cdot \boldsymbol{x})) \cdot \nabla_s L_f^-(s)\right].$$

The operator $\nabla_x \cdot \nabla_s$ appearing in (3.125,126) arises from the fact that the coordinates $\boldsymbol{x}$ and $\boldsymbol{s}$ are not independent. For an initial ground state, the amplitude $R_{if}^{CDW}(\boldsymbol{\eta})$ can be obtained, so that, [*Belkić* (1978)]

$$|R_{if}^{CDW}(\boldsymbol{\eta})|^2 = A(K, k) \, |N(\zeta)|^2 \, |N(\nu)|^2$$
$$\times |_2F_1(i\nu, i\zeta; 1; \tau) - i\nu \, w \, _2F_1(1 + i\nu, 1 + i\zeta; 2; \tau)|^2 \, |R_{if}^{B1}(\boldsymbol{\eta})|^2,$$

$$(3.127)$$

where

$$A(K, k) = \begin{cases} 1, & K^2 > k^2 + Z_t^2, \\ e^{-2\pi\nu}, & K^2 < k^2 + Z_t^2, \end{cases} \qquad (3.128)$$

$$w = \frac{\alpha}{\gamma} \frac{B\delta + C\gamma}{B(\alpha + \beta)}, \qquad (3.129)$$

$$B = K^2 - (1 + i\xi) \, \boldsymbol{K} \cdot \boldsymbol{k}, \qquad (3.130)$$

$$C = \frac{v}{p}\left[-\boldsymbol{K} \cdot \boldsymbol{k} + k^2(1 + i\xi)\right] + \left(1 + \frac{v}{p}\right)$$
$$\times \left[\left(\frac{k^2}{2} - \epsilon_i\right) - \boldsymbol{v} \cdot \boldsymbol{k}(1 + i\xi)\right], \qquad (3.131)$$

$$\alpha = \frac{K^2}{2}, \qquad \beta = -\boldsymbol{K} \cdot \boldsymbol{v}, \qquad \gamma = -\boldsymbol{K} \cdot \boldsymbol{p} + \alpha,$$

$$\delta = \boldsymbol{p} \cdot \boldsymbol{v} - pv + \beta, \qquad \tau = \frac{\beta\gamma - \alpha\delta}{\gamma(\alpha + \beta)}. \qquad (3.132)$$

It is noted that if we set $\nu = 0$, which corresponds to setting the initial distorted wave $\chi_i^{DWB} = \Phi_i^B$, we obtain the Distorted Wave Born (DWB) approximation, so that

$$\left| R_{if}^{DWB}(\boldsymbol{\eta}) \right|^2 = |N(\zeta)|^2 \left| R_{if}^{B1}(\boldsymbol{\eta}) \right|^2. \qquad (3.133)$$

The ECC peak is governed by the term $|N(\zeta)|^2$, in a form similar to Salin's theory. However, the factor ζ in (3.133) differs from the factor Z/p introduced by *Salin* (1969). It should be noted from (3.127–132) that in the CDW approximation the ECC peak shows a more complicated dependence with $\boldsymbol{p}$ than the one given by $|N(\zeta)|^2 \longrightarrow 2\pi Z_p/p$ as $p \to 0$, which corresponds to the DWB approximation.

3.5.2 Continuum Distorted Wave-Eikonal Initial State Approximation

Total cross sections obtained with the CDW approximation converge to those calculated with the B1 model as $v \to \infty$ [*Belkić* (1978)]. Hence, the CDW is expected to compare well with experiments at high energies. However, it significantly overestimates the experiments at intermediate impact velocities ($v \approx \langle k_i \rangle$). The same failure occurs for electron capture [*Crothers* (1982)], and has been attributed to the fact that the CDW initial Ansatz is not properly normalized [*Crothers* and *McCann* (1983)].

In order to solve this problem and to avoid the computational difficulties introduced by the normalization of the initial wave function, *Crothers* and *McCann* (1983), proposed to replace the distorting Coulomb factor, which describes the electron moving in a continuum state of the projectile, by its Eikonal approximation [*Fainstein* et al. (1987, 1988a, 1991), *Rivarola* et al. (1989)]. The replacement is supported by the fact that total cross sections are dominated by soft collisions. It is well known that both forms approach each other for $|vs + \boldsymbol{v} \cdot \boldsymbol{s}| \gg 1$. In particular, the approach occurs at sufficiently large distances R and at large impact parameters (separate encounters).

In this improved approximation, called Continuum Distorted Wave- Eikonal Initial State (CDW-EIS), the initial distorted wave function is thus chosen as

$$\chi_i^{EIS,+} = \Phi_i^B \, L_i^{EIS,+}(\boldsymbol{s}) = \Phi_i^B \, \exp\left[-\mathrm{i}\frac{Z_p}{v} \ln(vs + \boldsymbol{v} \cdot \boldsymbol{s}) \right]. \qquad (3.134)$$

It is obvious that this function verifies the asymptotic limit given by (3.121). The final distorted wave function is chosen as in the CDW approximation.

The perturbing potential in the entrance channel is now given by

$$(H - E)\chi_i^{EIS,+} = W_i \chi_i^{EIS,+} \tag{3.135}$$

$$= \Phi_i^B \left[\frac{1}{2} \nabla_s^2 L_i^{EIS,+}(s) + \nabla_x \ln \varphi_i(x) \cdot \nabla_s L_i^{EIS,+}(s) \right].$$

For the $1s$ ground state of a monoelectronic target, *Crothers* and *McCann* (1983) have obtained

$$|R_{if}^{CDW-EIS}(\boldsymbol{\eta})|^2 = |N(\zeta)|^2 |N(\nu)|^2 \exp[-2\pi\nu] \, |_2F_1(i\nu, i\zeta; 1; z)$$
$$-i\nu\Omega \, {}_2F_1(i\nu + 1, i\zeta + 1; 2; z)|^2 |R_{if}^{B1}(\boldsymbol{\eta})|^2 \tag{3.136}$$

with

$$\Omega = \alpha \left(B\delta + \gamma' C \right) / \gamma' \beta B, \tag{3.137}$$

$$C = \frac{v}{p} \left[-\boldsymbol{K} \cdot \boldsymbol{k} + k^2(1 + i\xi) \right] + \left(1 + \frac{v}{p} \right) [\Delta\epsilon - \boldsymbol{v} \cdot \boldsymbol{k}(1 + i\xi)], \tag{3.138}$$

$$\gamma' = \frac{1}{2}[Z_t^2 + (\boldsymbol{k} - \boldsymbol{K})^2], \tag{3.139}$$

and B, α, δ and β are defined as in the CDW approximation, and z is defined as

$$z = 1 - \frac{\alpha\delta}{\beta\gamma'}. \tag{3.140}$$

Thus, within the ECC peak, the CDW-EIS approximation also involves a complicated dependence on $\boldsymbol{p}$.

If the initial wave function is taken as $\chi_i^{EIS,+}$ and the final distorted wave function is obtained from $\chi_f^{CDW,-}$ approximating the projectile continuum factor $N(\zeta) \, {}_1F_1[-i\zeta; 1; -i(ps + \boldsymbol{p} \cdot \boldsymbol{s})]$ by its corresponding asymptotic eikonal phase, given by $\exp[-i(Z_p/p) \ln(ps + \boldsymbol{p} \cdot \boldsymbol{s})]$, the Symmetric Eikonal approximation for excitation (SEE) [*Fainstein* and *Rivarola* (1987)] is obtained. If the same choice is made for the initial state while in $\chi_f^{CDW;-}$ the target continuum factor $N(\zeta) \, {}_1F_1[-i\zeta; 1; -i(kx + \boldsymbol{k} \cdot \boldsymbol{x})]$ is replaced by the eikonal phase $\exp[-i(Z_t/k) \ln(kx + \boldsymbol{k} \cdot \boldsymbol{x})]$ the Symmetric Eikonal approximation for capture (SEC) is found [*Fainstein* and *Rivarola* (1987)] see also *Rivarola* and *Fainstein* (1987)]. These models for ionization can be considered as extensions of the symmetric Eikonal approximation introduced to describe electron capture, [*Maidagan* and *Rivarola* (1984)], and electron excitation, [*Deco* et al. (1986)] to bound states. The idea is to preserve the two center character of the final wave function favoring the projectile center in SEC or the target center in SEE.

Fainstein et al. (1987; 1988a,b; 1989a,b) have extended the CDW-EIS model to study single ionization from multi-electron targets by impact of bare projectiles. They reduced the problem to a one active electron reaction. The initial bound state was described by a Roothaan-Hartree-Fock function [*Clementi* and *Roetti* (1974)], given by the expression

$$\varphi_i(\boldsymbol{x}) = \sum_j C_j \, R_j(x) \, Y_{l_j}^{m_j}(\hat{\boldsymbol{x}}) \tag{3.141}$$

with

$$R_j(x) = [(2n_j)!]^{-1/2} \, (2Z_j)^{n_j+1/2} \, x^{n_j-1} \exp\left(-Z_j x\right) \tag{3.142}$$

being $n_j l_j m_j$ the quantum numbers corresponding to the Slater orbital $R_j(x) \, Y_{l_j}^{m_j}(\hat{\boldsymbol{x}})$. Then, the initial distorted wave function is chosen using (3.134).

In the exit channel, the Hartree-Fock target potential was approximated by a Coulomb potential $(-Z_t^*/x)$, where Z_t^* is an effective charge chosen in agreement with the prescription used by *Belkić* et al. (1979) for single electron capture. They have taken $Z_t^* = \left(-2n_i^2 \, \epsilon_i\right)^{1/2}$, with ϵ_i the initial electron orbital energy. The final distorted wave function now reads:

$$\chi_f^{CDW,-} = (2\pi)^{-3} \exp[\mathrm{i}\,\boldsymbol{K}_f \cdot \boldsymbol{R}_i + \mathrm{i}\,\boldsymbol{k} \cdot \boldsymbol{x}] \, N(\xi^*)$$
$$\times {}_1F_1\left(-\mathrm{i}\xi^*; 1; -\mathrm{i}(kx + \boldsymbol{k} \cdot \boldsymbol{x})\right) N(\zeta) \, {}_1F_1\left(-\mathrm{i}\zeta; 1; -\mathrm{i}(ps + \boldsymbol{p} \cdot \boldsymbol{s})\right) \tag{3.143}$$

with $\xi^* = Z_t^*/k$. Analytical expressions for the amplitude $R_{if}^{CDW-EIS}(\boldsymbol{\eta})$ have been recently calculated within the CDW-EIS model [*McCartney* and *Crothers* (1993)].

The CDW-EIS model has been employed with success to describe double-differential, single differential and total cross sections to study numerous systems [*Fainstein* et al. (1991)]. They include the impact of multiply charged heavy ions, protons and antiprotons as projectiles.

One important limitation in the previous model lies in the description of the continuum target state in the exit channel. Like in the B1 approximation it underestimates the double differential cross sections at backward emission angles. In order to improve the model, different authors [*Fainstein* et al. (1994), *Schultz* and *Reinhold* (1994), *Gulyás* et al. (1995a)] have introduced Hartree-Fock numerical bound and continuum target states in the CDW-EIS model as it has been done by *Madison* (1973) in the case of the B1 approximation. The initial bound state $\varphi_i(\boldsymbol{x})$ is taken as in (3.103) and $\varphi_f^-(\boldsymbol{x})$ is expanded in partial waves as in (3.101). The resulting amplitude reads [*Gulyás* et al. (1995a,b)]:

$$R_{if}^{CDW-EIS}(\boldsymbol{\eta}) = \frac{4\pi}{v} Z_p \frac{N(\zeta)\Gamma(1 - \mathrm{i}\nu)}{\beta} \left(\frac{\alpha}{\gamma}\right)^{\mathrm{i}\zeta} \left(\frac{\alpha}{\mathrm{i}\beta}\right)^{\mathrm{i}\nu}$$
$$\times \sum_{l'm'} (-\mathrm{i})^{l'} \mathrm{e}^{\mathrm{i}\delta_i'} \left[Y_{l'}^{m'}(\hat{\boldsymbol{k}})\right]^* \mathrm{e}^{\mathrm{i}M\phi_Q} \{A_1 C_{l'm'}$$
$$+ A_2 \left[-\frac{\Delta\epsilon}{v} L_z^0 + \eta \left(L_x^+ + L_x^-\right)\right] \tag{3.144}$$
$$+ A_3 \left(k_{//} - v - p\right) L_z^0 + A_3 \, k_\perp \left[L_x^+ \, \mathrm{e}^{\mathrm{i}\varPhi} + L_x^- \, \mathrm{e}^{-\mathrm{i}\varPhi}\right]\},$$

where $\delta_{l'}$ is the phase shift, $M = m' + m_i$, $k_{//} = \boldsymbol{k} \cdot \hat{\boldsymbol{v}}$, $k_\perp = \boldsymbol{k} \cdot \hat{\boldsymbol{\eta}}$, $\boldsymbol{Q} = -\boldsymbol{K}$, $\Phi = \phi_Q - \phi_e$, with ϕ_Q and ϕ_e the azimuthal angles subtended by $\boldsymbol{Q}$ and $\boldsymbol{k}$, respectively,

$$A_1 = -\mathrm{i}\left[F(1) + \mathrm{i}\zeta\, z\, F(2)\right], \tag{3.145}$$

$$A_2 = \frac{1}{\alpha}\left[F(1) - \mathrm{i}\zeta\,(1-z)\frac{\delta-\beta}{\delta}F(2)\right], \tag{3.146}$$

$$A_3 = \frac{\mathrm{i}\zeta\,(1-z)}{\alpha\,\delta}\beta\,F(2), \tag{3.147}$$

with $F(1) = {}_2F_1(\mathrm{i}\zeta, \mathrm{i}\nu; 1; z)$, $F(2) = {}_2F_1(1+\mathrm{i}\zeta, 1+\mathrm{i}\nu; 2; z)$ and

$$
\begin{aligned}
C_{l'm'} = \sum_l (-\mathrm{i})^l(-1)^M &\sqrt{\frac{(2l+1)(2l'+1)(2l_i+1)}{4\pi}} \\
\times &\begin{pmatrix} l & l' & l_i \\ 0 & 0 & 0 \end{pmatrix}\begin{pmatrix} l & l' & l_i \\ -M & m' & m_i \end{pmatrix} \\
\times &\mathcal{Y}_l^M(\hat{\boldsymbol{Q}})\int dx\; u_{n_i l_i}(x)j_l(Qx)u_{\epsilon l'}(x),
\end{aligned}
\tag{3.148}
$$

$$L_x^\pm(\boldsymbol{Q},\epsilon) = \pm\frac{1}{\sqrt{2}}\left(A_{l'm'}^{l_i^+ m_i^\pm} + B_{l'm'}^{l_i^- m_i^\pm}\right), \tag{3.149}$$

$$L_z^0(\boldsymbol{Q},\epsilon) = A_{l'm'}^{l_i^+ m_i} + B_{l'm'}^{l_i^- m_i}, \tag{3.150}$$

with $\mathcal{Y}_l^m = Y_l^m \exp(-\mathrm{i}m\phi)$, $l_i^\pm = l_i \pm 1$, $m_i^\pm = m_i \pm 1$, $M^\pm = M \pm 1$. The functions $L_x^\pm$ and L_z^0 require the calculation of Fourier-Bessel transforms including the derivative of $u_{n_i l_i}(x)$:

$$
\begin{aligned}
A_{l'm'}^{\lambda_1\lambda_2} = \sum_l (-\mathrm{i})^l(l_i+1)^{1/2}\mathcal{Y}_l^m(\hat{\boldsymbol{Q}})&\begin{bmatrix} l & l' & \lambda_1 \\ m & m' & \lambda_2 \end{bmatrix} \\
\times &\int dx\frac{xu'_{n_i l_i}(x) - (l_i+1)u_{n_i l_i}(x)}{x}j_l(Qx)u_{\epsilon l'}(x)
\end{aligned}
\tag{3.151}
$$

and

$$
\begin{aligned}
B_{l'm'}^{\lambda_1\lambda_2} = -\sum_l (-\mathrm{i})^l l_i^{1/2}\mathcal{Y}_l^m(\hat{\boldsymbol{Q}})&\begin{bmatrix} l & l' & \lambda_1 \\ m & m' & \lambda_2 \end{bmatrix} \\
\times &\int dx\frac{xu'_{n_i l_i}(x) + l_i u_{n_i l_i}(x)}{x}j_l(Qx)u_{\epsilon l'}(x),
\end{aligned}
\tag{3.152}
$$

where the angular coefficients are:

$$
\begin{aligned}
\begin{bmatrix} l & l' & \lambda_1 \\ m & m' & \lambda_2 \end{bmatrix} = &\sqrt{\frac{(2l+1)(2l'+1)(2\lambda_1+1)}{4\pi}} \\
\times &\begin{pmatrix} \lambda_1 & 1 & l_i \\ \lambda_2 & m_i - \lambda_2 & -m_i \end{pmatrix}\begin{pmatrix} l & l' & \lambda_1 \\ 0 & 0 & 0 \end{pmatrix} \\
\times &\begin{pmatrix} l & l' & \lambda_1 \\ -m & m' & \lambda_2 \end{pmatrix}.
\end{aligned}
\tag{3.153}
$$

The radial functions $u_{n_i l_i}$ and $u_{\epsilon l}$ are calculated by a numerical integration of the radial equation. The Fourier-Bessel transforms (3.148,151,152) are then evaluated numerically.

Extensive calculations of double-differential cross sections for impact of different bare ions on He, Ne, and Ar targets at intermediate and high energies, show improvement describing experimental data at small and large electron angles over a wide range of electron energies [*Fainstein* et al. (1994), *Gulyás* et al. (1995a,b), *Stolterfoht* et al. (1995)]. Improvement is also observed for the case of total cross sections.

Calculations of CDW-EIS double differential cross sections recently developed using the same model potential [*Garvey* et al. (1975)] to describe the initial bound and final continuum target states for ionization of He [*Schultz and Reinhold* (1994)], are in agreement with those presented by *Gulyás* et al. (1995a). However, in the TCEE region, CDW-EIS underestimates experimental data for zero-degree electron ejection for energetic F^{9+} impact, Sect. 5.3.1.

Finally, we note that if the initial and final wave functions are chosen as

$$\chi_i^+ = \chi_i^{EIS,+} \tag{3.154}$$

and

$$\chi_f^- = \Phi_f^B \tag{3.155}$$

the eikonal approximation is obtained. These initial and final wave functions are equivalent to those used to describe the Glauber approximation [*Glauber* (1959) *McGuire* (1982, 1983)].

3.5.3 Quantum-Mechanical Impulse Approximation

In a recent work, *Miraglia* and *Macek* (1991) have introduced an impulsive quantum-mechanical approximation to describe electron ionization. These authors chose the final distorted wave function $\chi_f^{CDW,-}$ to describe the exit channel. The difference with the previous distorted wave models lies in the choice of the initial wave function. The entrance channel is described by an impulse wave function, so that the method is named *impulse approximation* (IA). This model is expected to be valid for $Z_p > v > Z_t/n$, with n the principal quantum of the bound state.

The initial state in the B1 approximation can be written as

$$\Phi_i^B = (2\pi)^{-3} \int d\boldsymbol{k}_i \tilde{\varphi}_i(\boldsymbol{k}_i) \exp\left(i\boldsymbol{K}_{\mathrm{i}} \cdot \boldsymbol{R}_{\mathrm{i}}\right) \exp\left(i\boldsymbol{k}_i \cdot \boldsymbol{x}\right)$$

$$= (2\pi)^{-3} \int d\boldsymbol{k}_i \tilde{\varphi}_i(\boldsymbol{k}_i) \exp[i(\mu_T(\boldsymbol{K}_i + \boldsymbol{k}_i) \cdot \boldsymbol{R}_f + i\mu_P(\boldsymbol{k}_i - \boldsymbol{v}) \cdot \boldsymbol{s}], \tag{3.156}$$

where the integrand in the first line is written as a function of the initial Jacobi coordinates $(\boldsymbol{x}, \boldsymbol{R}_i)$ and the integral in the second line as a function of the Jacobi coordinates $(\boldsymbol{s}, \boldsymbol{R}_f)$. The vector $\boldsymbol{R}_f$ gives the position of the center of mass of the projectile-electron sub-system with respect to the target nucleus, and $\mu_P = M_P/(M_P + 1)$ is the reduced mass corresponding to this sub-system. The term $\exp\left[i\mu_P\left(\boldsymbol{k}_i - \boldsymbol{v}\right) \cdot \boldsymbol{s}\right]$ shows that the electron motion is described by a plane wave with the momentum $(\boldsymbol{k}_i - \boldsymbol{v})$ in the projectile frame of reference.

The key of the impulse approximation consists in replacing this plane wave by a Coulomb continuum wave function $\psi^+(Z_p, \mu_P(\boldsymbol{k}_i - \boldsymbol{v}), \boldsymbol{s})$, where

$$\begin{aligned}
\psi^+(Z, \boldsymbol{k}, \boldsymbol{r}) = {}& (2\pi)^{-3/2} \exp\left(i\boldsymbol{k} \cdot \boldsymbol{r}\right) \ N^*(Z/k) \\
& \times {}_1F_1\left(iZ/k; 1; i(kr - \boldsymbol{k} \cdot \boldsymbol{r})\right)
\end{aligned} \tag{3.157}$$

so that

$$\begin{aligned}
\Phi_i^{IA} = {}& (2\pi)^{-3/2} \int d\boldsymbol{k}_i \ \tilde{\varphi}_i(\boldsymbol{k}_i) \\
& \times \exp[(\mu_T \boldsymbol{K}_i + \boldsymbol{k}_i) \cdot \boldsymbol{R}_f] \ \psi^+ \left[Z_p, \mu_P(\boldsymbol{k}_i - \boldsymbol{v}), \boldsymbol{s}\right].
\end{aligned} \tag{3.158}$$

The goal of this choice is the inclusion of the potential $(-Z_p/s)$ in all orders in the initial wave function. The term $\exp\left[i\left(\mu_T \boldsymbol{K}_i + \boldsymbol{k}_i\right) \cdot \boldsymbol{R}_f\right]$ describes the relative movement between the projectile-electron sub-system and the target nucleus. According to *Coleman* (1969) in the impulse hypothesis the target electron is regarded as free during the collision, but its momentum distribution is taken to be that of the appropriate bound state. The wave function Φ_i^{IA} is thus a wave packet, describing the electron in a continuum state of the projectile, modulated by the initial distribution of the electron, $\tilde{\varphi}_i(\boldsymbol{k}_i)$. The corresponding perturbing potential results from

$$\begin{aligned}
(H - E)\Phi_i^{IA} = {}& (2\pi)^{-3/2} \int d\boldsymbol{k}_i \tilde{\varphi}_i(\boldsymbol{k}_i) \left(\frac{\mu_P k_i^2}{2} - \epsilon_i - \frac{Z_t}{x}\right) \\
& \times \exp[i(\mu_T \boldsymbol{K}_i + \boldsymbol{k}_i) \cdot \boldsymbol{R}_f] \ \Psi^+ \left(Z_p, \mu_P(\boldsymbol{k}_i - \boldsymbol{v}), \boldsymbol{s}\right).
\end{aligned} \tag{3.159}$$

After some algebra using the Nordsiek technique, the T_{if}^{IA} - matrix element is given by the integral expression:

$$\begin{aligned}
T_{if}^{IA} = {}& -\frac{1}{\pi^4} \int d\boldsymbol{k}_i \ \tilde{\varphi}_i(\boldsymbol{k}_i) \ N^*(\xi) \ N^*(\zeta) \ N^*(Z_p/p_i) \\
& \times A_P^{-i\zeta-1} A_T^{-i\xi-1} A_i^{-Z_p/p_i} \frac{Z_p Z_t}{K^4} \frac{1}{S^4} \Bigg\{ {}_2F_1\left(iZ_p/p_i, i\zeta; 1; X\right)\boldsymbol{K} \\
& + i\frac{Z_p}{p_i A_i} \left(\boldsymbol{p}_i + p_i \hat{\boldsymbol{p}} + \frac{A_3 - A_i}{A_P}\boldsymbol{K}\right) \\
& \times {}_2F_1[i(Z_p/p_i) + 1, i\zeta + 1; 2; X] \Bigg\} \cdot \boldsymbol{S},
\end{aligned} \tag{3.160}$$

where

$$\boldsymbol{p}_i = \boldsymbol{k}_i - \boldsymbol{v}, \quad \boldsymbol{S} = \boldsymbol{K} - \boldsymbol{k} + \boldsymbol{k}_i, \tag{3.161}$$

$$A_T = 1 + 2\frac{\boldsymbol{S} \cdot \boldsymbol{k}}{S^2}, \ A_P = 1 - 2\frac{\boldsymbol{K} \cdot \boldsymbol{p}}{K^2}, \tag{3.162}$$

$$A_i = 1 + 2\frac{\boldsymbol{K} \cdot \boldsymbol{p}_i}{K^2}, \ A_3 = 1 + \frac{2\left(p_i p + \boldsymbol{p}_i \cdot \boldsymbol{p}\right)}{K^2},$$

$$X = 1 - \frac{A_P - A_i - A_3}{A_P A_i}, \tag{3.163}$$

and the other quantities are defined as for CDW-EIS. If $Z_p = 0$ in (3.158), the first-order multiple scattering approximation is obtained [*Garibotti* and *Miraglia* (1980, 1981)]. If $(\boldsymbol{k}_i - \boldsymbol{v})$ is replaced by $(\boldsymbol{K} - \boldsymbol{v} - \boldsymbol{k})$ in (3.158) the modified CDW approximation [*Miraglia* (1983)] is reached.

The impulse approximation [*Miraglia* and *Macek* (1991)], is found to be in better agreement than CDW-EIS to reproduce double-differential cross sections for the impact of H^+ and He^{2+} ions on He targets when $k < v$, with $v = 2$ a.u. This agreement is also extended to the asymmetry of the ECC peak. The IA fails to predict the position of the maximum of the BE peak at $\theta = 0°$ for large Z_p (with $Z_p \gg Z_t$). This failure is due to treating the electron in a projectile continuum state from the start of the collision. However, the ionization event is mainly produced at a particularly large separation of the collision partners [*Bohr* and *Lindhard* (1954), *Olson* and *Salop* (1976), *Grozdanov* and *Janev* (1978)].

It has been shown that the position of the BE peak is influenced by this specific distance [*Pedersen* et al. (1990, 1991), *Fainstein* et al. (1992)]. As compared to the classical prediction, the BE peak is shifted to lower electron energies as the projectile charge increases. Note that this is a two-center effect which will be described in detail further on in the corresponding experimental section. At this point we only would like to emphasize that the CDW-EIS model gives an adequate description of the position of this peak. In the IA, each k-component of the initial momentum distribution is individually distorted from the beginning of the collision. On the contrary, in the CDW-EIS the whole spatial distribution of the initial bound state is distorted by the same average continuum factor, thus preserving this distribution at smaller internuclear distances.

If the function $\psi^+ (Z_p, \mu_P (\boldsymbol{k}_i - \boldsymbol{v}), \boldsymbol{s})$ is approximated by $\psi^+ (Z_p, -\mu_P \boldsymbol{v}, \boldsymbol{s}) \times \exp[\mathrm{i}\mu_p \boldsymbol{k}_i \cdot \boldsymbol{s}]$ in the evaluation of integral (3.158), using the argument that $\tilde{\varphi}_i(\boldsymbol{k}_i)$ peaks at $k_i = 0$, the initial distorted wave function Φ_i^{IA} coincides with $\chi_i^{CDW,+}$. However, CDW-EIS and IA should be considered as different approximations.

The IA was first introduced by *Chew* (1950), see also *Chew* and *Goldberg* (1952), to deduce the free neutron-neutron scattering cross section from measurements of n-d inelastic cross sections. Applications to atomic collision problems were given by *Pradhan* (1957), *McDowell* (1961), and *Coleman* (1968), among others. For ionization of the atomic target, the initial wave function was chosen in the form given by (3.158) but different final wave functions were proposed. So, the electron in the exit channel has been considered as a plane wave Φ_f^{PW} [*Akerib* and *Blorowitz* (1961)], a target continuum state Φ_f^B [*Coleman* (1969)], or a projectile continuum state Φ_f^{QBK} [*Jakubaßa-Amundsen* (1983, 1988), *Oswald* et al., 1994)]. All these models may be considered as particular cases of the generalized IA developed by *Miraglia* and *Macek* (1991).

3.5.4 Distorted Strong Potential Born Approximation and Related Models

As is well known the IA uses on-the-energy-shell intermediate states. A related approximation which introduces off-the-energy-shell intermediate states is the Distorted Strong Potential Born (DSPB), *Brauner* and *Macek* (1992), *Macek* and *Taulbjerg* (1993). The exact T-matrix element in a distorted wave approximation can be written as

$$T_{if} = \left\langle \chi_f^- \left| (V_T - U_f) \left[1 + G^+ (V_P - U_i) \right] \right| \chi_i^+ \right\rangle, \tag{3.164}$$

where G^+ is the full Green operator. In the previous equation, U_i and U_f are the initial and final distortion potentials, respectively. The distorted functions χ_i^+ and χ_f^- are determined correspondingly to these potentials. The DSPB approximation in asymmetric ion-atom collisions is defined by ignoring the weaker of the two perturbation potentials in the full Green operator. In the case $Z_p \gg Z_t$, the operator G^+ is approximated as

$$G^+ \cong G_P^+ = [E - (H - V_T) + i\eta]^{-1}. \tag{3.165}$$

A DSPB initial wave function is thus defined by the equation [*Taulbjerg* (1990a)]

$$\Phi_i^{DSPB} = \left[1 + G_P^+ (V_P - U_i) \right] \left| \chi_i^+ \right\rangle, \tag{3.166}$$

where the potential U_i is chosen as

$$U_i = \int d\boldsymbol{x} \, |\varphi_i(\boldsymbol{x})|^2 \, V_P(\boldsymbol{s}). \tag{3.167}$$

Then, using the near-shell approximation [*Taulbjerg* et al. (1990), *Taulbjerg* (1990a)] the DSPB-initial wave function takes the form

$$\Phi_i^{DSPB} = (2\pi)^{-3/2} \int d\boldsymbol{k}_i \, \tilde{\varphi}_i(\boldsymbol{k}_i) \, \exp[i(\mu_T \boldsymbol{K}_i + \boldsymbol{k}_i) \cdot \boldsymbol{R}_f]$$
$$\times \, \gamma^+(\boldsymbol{k}_i) \, \psi^+(Z_p, \mu_P(\boldsymbol{k}_i - \boldsymbol{v}), \boldsymbol{s}). \tag{3.168}$$

The factor $\gamma^+(\boldsymbol{k}_i)$, which contains the off-energy-shell character of the intermediate continuum states, is given to a good approximation [*Taulbjerg* (1990a)] by the expression

$$\gamma^+(\boldsymbol{k}_i) = i \int_0^\infty dx \, \exp(-ix) \, \exp\left\{ -\frac{i}{v} \int_{x/\kappa} \left[U_i(R) + \frac{Z_p}{R} \right] dR \right\} \tag{3.169}$$

with

$$\kappa = \frac{1}{v}\left(\frac{k_i^2}{2} - \epsilon_i\right). \tag{3.170}$$

The only difference between the wave functions Φ_i^{IA}, (3.158), and Φ_i^{DSPB} is the factor $\gamma^+(k_i)$. If in Φ_i^{DSPB}, $\gamma^+(k_i)$ is replaced by unity, we recover Φ_i^{IA} that contains on-the energy-shell intermediate states. Furthermore, in the DSPB calculations developed by *Brauner* and *Macek* (1992) the final wave function χ_f^- is chosen as $\chi_f^- = \Phi_f^{OBK}$ ($U_f = 0$), which yields a model similar to the IA introduced by *Jakuba\ss a-Amundsen* (1983).

It should also be noted that when the distortion potential U_i is ignored in Φ_i^{DSPB}, this wave function reduces to the usual Strong Potential Born (SPB) wave function Φ_i^{SPB} [*Macek* and *Shakeshaft* (1980), *Macek* and *Taulbjerg* (1981), *Taulbjerg* et al. (1990)].

In a recent work, *Madsen* and *Taulbjerg* (1995) have introduced the channel-distorted two-center (D2C) approximation. They have used a two-center wave function $\chi_f^- = \chi_f^{CDW,-}$ to describe the exit channel and the initial distorted wave function χ_i^D corresponding to the DSPB approximation:

$$\chi_i^D = \Phi_i^B \, D_v^+(\mathbf{R}), \tag{3.171}$$

where the multiplicative distortion factor is given by

$$D_v^+(\mathbf{R}) = (vR - \mathbf{v} \cdot \mathbf{R})^{-iv} \exp\left[-\frac{i}{v} \int\limits_{-\infty}^{R_Z} dR_z' \left(U_i(R') + \frac{Z_p}{R'}\right)\right] \tag{3.172}$$

with $R_Z = \mathbf{R} \cdot \hat{v}$. Considering only the first term of (3.166) they have calculated the transition matrix element (3.164) as

$$T_{if}^{D2C} = \langle \chi_f^{CDW,-} | (V_P - U_i) | \chi_i^D \rangle. \tag{3.173}$$

After some algebra T_{if}^{D2C} can be written as

$$T_{if}^{D2C}(\mathbf{K}) = (2\pi)^{-3/2} \int d\mathbf{J}\, \tilde{D}_v^+(\mathbf{J}) \left[\tilde{\psi}_p^{Z_p-}(\mathbf{p} - \mathbf{K} - \mathbf{J})\right]^*$$

$$\times \left[\epsilon_i - \frac{1}{2}(\mathbf{k} - \mathbf{K} - \mathbf{J})^2\right] F_{ok}(-\mathbf{K} - \mathbf{J}), \tag{3.174}$$

where $\tilde{D}_v^+$ and $\tilde{\psi}_p^{Z_p-}$ are the Fourier transforms of D_v^+ and $\psi^-(Z_p, \mathbf{p}, \mathbf{s})$, respectively, being $\psi^-(Z_p, \mathbf{p}, \mathbf{s})$ and $\psi^-(Z_t, \mathbf{k}, \mathbf{x})$ the incoming Coulomb waves of momentum $\mathbf{p}$ and $\mathbf{k}$ in the projectile and target fields, respectively. The function $F_{ok}(\mathbf{q})$ is the form factor for target ionization given by

$$F_{ok}(\mathbf{q}) = \langle \psi^-(Z_t, \mathbf{k}, \mathbf{x}) | \exp(-i\mathbf{q} \cdot \mathbf{x}) | \varphi_i(\mathbf{x}) \rangle. \tag{3.175}$$

Developing the Coulomb continuum wave function in momentum space $\tilde{\psi}_p^{Z_p-}(\mathbf{q})$ in its on-energy-shell and off-energy-shell components [*van Haering*

(1985)] and using a peaking approximation based on the behavior of $D_v^+(\boldsymbol{J})$ at $\boldsymbol{J} = \boldsymbol{0}$, *Madsen* and *Taulbjerg* (1995) have shown that

$$T_{if}^{D2C} \cong T^R(\boldsymbol{p}, \boldsymbol{p} - \boldsymbol{K}) \varsigma_p^+(\boldsymbol{p} - \boldsymbol{K}) F_{ok}(-\boldsymbol{K}), \qquad (3.176)$$

where $T^R(\boldsymbol{p}, \boldsymbol{q})$ is the off-energy-shell transition matrix element for the Rutherford scattering of the electron in the projectile field:

$$T^R(\boldsymbol{p}, \boldsymbol{q}) = -\frac{1}{2\pi^2} \frac{Z_p}{|\boldsymbol{p} - \boldsymbol{q}|^2} \left(\frac{|\boldsymbol{p} - \boldsymbol{q}|^2}{(p+q)^2} \right)^{i\zeta} \frac{\Gamma(1 - i\zeta)}{\Gamma(1 + i\zeta)} \qquad (3.177)$$

with $\zeta = Z_p/p$ and

$$\varsigma_p^+(\boldsymbol{q}) = (2\pi)^{-3/2} \int d\boldsymbol{S} \, \tilde{D}_v^+(\boldsymbol{S}) \frac{p^2 - q^2}{p^2 - (\boldsymbol{q} - \boldsymbol{S})^2} g_p^+(|\boldsymbol{q} - \boldsymbol{S}|), \quad (3.178)$$

where $g_p^+(q)$ is the Coulomb off-shell factor given by

$$g_p^+(q) = \Gamma(1 + i\zeta) \exp\left(\frac{\pi Z_P}{2p} \right) \left(\frac{q - p}{q + p} \right)^{-i\zeta}. \qquad (3.179)$$

The factored form (3.176) is accurate for final electron velocities of the order of or larger than the beam velocity, which include the ECC and BE regions. It has been shown that the factor ς_p^+ is almost identical to the off-energy-shell factor $\gamma^+(k_i)$ appearing in the DSPB approximation [*Madsen* and *Taulbjerg* (1994)].

The matrix element $T^R(\boldsymbol{p}, \boldsymbol{p} - \boldsymbol{K})$ corresponds to the scattering of an electron with initial momentum $\boldsymbol{p} - \boldsymbol{K} = \boldsymbol{k}_i - \boldsymbol{v}$ and final momentum $\boldsymbol{p}$, with respect to the projectile nucleus. Moreover, using (3.176)

$$\left| T_{if}^{D2C}(\boldsymbol{K}) \right|^2 \cong \left| T_{if}^{B1}(\boldsymbol{K}) \right|^2 \left| \varsigma_p^+(\boldsymbol{p} - \boldsymbol{K}) \right|^2. \qquad (3.180)$$

The factor ς_p^+ approaches unity in the near shell limit, so that T_{if}^{D2C} reduces to the B1 approximation in the BE peak. However, away from the BE peak, ς_p^+ is a strongly varying function of the electron momentum [*Madsen* and *Taulbjerg* (1994)]. Doubly-differential cross sections for impact of F^{9+} on He targets at a collision energy of $1.5\,\mathrm{MeV/u}$ calculated with the D2C model have been shown to be in good agreement with experimental data at forward electron scattering angles.

If a further approximation is used in D2C, describing the final distorted wave function as in the OBK1 model, the distorted Brinkman-Kramers (DBK) approximation is obtained [*Madsen* and *Taulbjerg* (1994, 1995)]. Following the same procedure as the one used for the evaluation of T_{if}^{D2C}, the scattering matrix element corresponding to the DBK results in the factored form:

$$T_{if}^{DBK} = T^R(\boldsymbol{p}, \boldsymbol{p} - \boldsymbol{K}) \varsigma_p^+(\boldsymbol{p} - \boldsymbol{K}) \tilde{\varphi}_i(\boldsymbol{k}_i). \qquad (3.181)$$

It should be noted that the D2C matrix element reduces to the DBK one, replacing the form factor $F_{ok}(-\boldsymbol{K})$ by the Fourier transform of the initial bound wave function $\tilde{\varphi}_i(\boldsymbol{k}_i)$ in (3.176). It has been shown that the DBK approximation corrects the inadequate representation of the BE peak given by the OBK approximation [*Madsen* and *Taulbjerg* (1994)]. Moreover, the DBK approximation becomes similar to the elastic scattering model (Sect. 3.1.2) when $\varsigma_p^+ \cong 1$, as was the case in the BE peak.

3.6 Close-Coupling Calculations

The impact parameter approximation is frequently used to describe ion-atom collisions [*McDowell* and *Coleman* (1970)]. In this approximation the projectile is assumed to move along a classical trajectory given by a vector $R(\boldsymbol{b}, t)$ where $\boldsymbol{b}$ is the impact parameter. The electron wave function $\Psi(t)$ must solve the time-dependent Schrödinger equation:

$$\left(H_T + V_P - \mathrm{i}\frac{d}{dt}\right)\Psi\left(t\right) = 0, \tag{3.182}$$

where H_T is the Hamiltonian of the atomic target and V_P is the projectile-electron interaction.

An approximation commonly used in the evaluation of (3.182) consists of assuming that the projectile moves with constant velocity $\boldsymbol{v}$, so that $\boldsymbol{R} = \boldsymbol{b} + \boldsymbol{v}t$. This approximation is known as the straight line version of the impact parameter method. Other projectile paths have been considered using the Newton's classical laws of motion where the projectile has been described moving in an effective field $W(\boldsymbol{R})$. In the case where only the internuclear potential is taken into account the projectile deflects in hyperbolic trajectories [*Martir* et al. (1982)]. *Schiwietz* (1990) has considered that the projectile and electronic motions are coupled assuming that the heavy particle moves in a potential $W(\boldsymbol{R}) = Z_p Z_t/R + \langle \Psi | V_P(\boldsymbol{s}) | \Psi \rangle$. It has also been shown that projectile deflections can affect the ionization cross sections at low enough impact energies [*Martir* et al. (1982), *Schiwietz* (1990)].

In the close coupling method, the wave function $\Psi(t)$ is expanded as a linear combination of a complete basis of eigenstates of the Hamiltonian $H = H_T + V_P$. However, in order to calculate $\Psi(t)$, the basis of Eigenstates must be truncated and the wave function can only be determined in an approximate form. Moreover, a difficult task arises when continuum states are explicitly incorporated in the basis. In general, the trial function $\Psi_{tr}(t)$ is expanded in a finite set of pseudo-eigenstates $\chi_k(t)$,

$$\Psi_{tr}(t) = \sum_{k=1}^{N} a_k(t)\,\chi_k(t), \tag{3.183}$$

where $|a_k(t)|^2$ gives the probability to occupy the pseudo-state $\chi_k(t)$ as $t \to \infty$.

The pseudo-states may be chosen as linear combinations of a finite set of square integrable functions. Thus, the infinite Hilbert space is divided in a finite P-subspace defined by the projector,

$$P = \sum_{k=1}^{N} |\chi_k\rangle \langle \chi_k|, \qquad (3.184)$$

and in its complementary Q-subspace, so that the corresponding projector operators are subject to the equations $P + Q = 1$ and $PQ = 0$. Thus, orthonormal pseudo-states $\chi_k(t)$ are eigenvectors of the operator PHP and consequently, diagonalize it. The coefficients $a_k(t)$ can be obtained solving the equation

$$\left\langle \chi_k \left| H - i\frac{d}{dt} \right| \Psi_{tr} \right\rangle = 0 \qquad (3.185)$$

which will generate a system of coupled equations.

Inner-shell ionization has been studied using a single-center expansion (SCE) [*Reading* et al. (1976), *Ford* et al. (1977), *Swafford* et al. (1977) *Reading* and *Ford* (1979), *Reading* et al. (1979)], where the pseudo-states $\chi_k(t)$ are chosen to diagonalize the operator PH_TP :

$$\langle \chi_{k'} | H_T | \chi_k \rangle = \lambda_k \delta_{k'k}. \qquad (3.186)$$

A three-state SCE has been employed by *Janev* and *Presnyakov* (1980) including an initial ns state, an intermediate $n'p$ $(m = 0)$ state and an effective continuum ϵp state. The potential V_P is described in its dipolar approximation. Using the concept of the effective oscillator strength for the transitions into the continuum, the continuum spectrum is replaced by an effective energy level. The model has been used with some success to describe total ionization cross sections [*Fite* et al. (1960), *Gilbody* and *Ireland* (1964) *Park* et al. (1977)] in $H^+ + H$ collisions. However, this close-coupling approximation systematically underestimates the experiments, especially at the cross section maximum. This has been attributed to the exclusion of multi-step transitions to the continuum through discrete intermediate states. Total cross sections have been also determined for impact of multicharged bare ions on H atoms.

However, as the electron capture channels become relevant to determine the electron flux loss from the target, as is the case when the projectile nuclear charge increases at low and intermediate impact energies, a huge basis χ_k is necessary to describe the collision. Only when $N \to \infty$, the set χ_k subtends the infinite Hilbert space of solutions of (3.182) and the charge exchange channels will be completely contained in $\Psi_{tr}(t)$ at all values of t .

In order to give a practical solution to this limitation of the SCE model and following the approach of *Gallaher* and *Wilets* (1968), *Shakeshaft* (1976, 1978) introduced a two-center expansion (TCE) to study the $H^+ + H$ collision system. In this system, the flux of probability towards the resonant electron capture channels influences the population of target bound and free states. Shakeshaft has considered the finite basis of Sturmian [*Shakeshaft* (1976)] or scaled hydrogenic [*Shakeshaft* (1978)] functions to generate the pseudo-states $\chi_k(t)$. Using the symmetry properties of the homonuclear system, the pseudo-states $\chi_k(t)$ which diagonalize operators of the forms PH_TP and PH_PP are determined, H_P being the Hamiltonian that corresponds to the projectile channels with perturbing potential V_T. These Sturmian and scaled hydrogenic expansions form discrete sets which are overcomplete.

To obtain the total ionization cross section, the probability of occupying a particular pseudo-state $\chi_k(t)$ at the end of the collision is multiplied by the square of the overlap matrix element of this eigenvector with the continuum of the hydrogen atom whose nucleus also provides the center for $\chi_k(t)$. Then, the ionization probability is obtained by summing these products over all eigenvectors. This procedure does not result in double-counting because projectile and target eigenvectors are linearly independent at the end of the collision.

Ionization in $H^+ + H$ collisions has been also treated using a Gaussian basis [*Dose* and *Semini* (1974), *Toshima* and *Eichler* (1992), *Toshima* (1994)]. We should also mention that Winter has extended the coupled-Sturmian-pseudo-state model for collisions of protons with arbitrary hydrogenic-ion targets [*Winter* (1982, 1986, 1987), *Stodden* et al. (1990), *Winter* and *Alston* (1992)] and neutral targets [*Winter* (1991, 1993)].

In principle, the TCE model can be systematically improved by increasing the number of target and projectile pseudo-states. However, doing so requires extensive computer time. Different modifications to accelerate the calculations have been proposed by *Reading* et al. (1981), for applications see *Martir* et al. (1982) and *Ford* et al. (1982).

A possible solution to the study of ionization from the target is obtained through the inclusion of optical potentials [*Lüdde* et al. (1987, 1988)]. For simplicity we consider the case in which the P-space includes all important discrete excitation channels of the target, and its complementary Q- space includes all the other possible states $\alpha_k(t)$ populated during the collision,

$$Q = \sum_{k=N+1}^{\infty} |\alpha_k\rangle \langle \alpha_k| . \qquad (3.187)$$

Then, the system of coupled equations to be solved are

$$i\dot{a}_k = \sum_{j=1}^{N} a_j \langle \chi_k| V_P + \epsilon_j |\chi_j\rangle + \sum_{j=N+1}^{\infty} b_j \langle \chi_k| V_P |\alpha_j\rangle , \qquad (3.188)$$

$$i\dot{b}_k = \sum_{j=1}^{N} a_j \langle \alpha_k| V_P |\chi_j\rangle + \sum_{j=N+1}^{\infty} b_j \langle \alpha_k| V_P + \epsilon_j |\alpha_j\rangle , \qquad (3.189)$$

where $b_k(t)$ is the coefficient corresponding to the function $\alpha_k(t)$ in a linear combination of the electronic wave function over the infinite complete space. Solving (3.189) in terms of a perturbation series and replacing the result in (3.188), the following equations are obtained for the coefficients corresponding to the P-space:

$$i\dot{a}_k = \sum_{j=1}^{N} a_j \langle \chi_k | \, V_P + \epsilon_j \, | \chi_j \rangle + \sum_{j=1}^{N} \langle \chi_k | \, V_P v_j \, | \chi_j \rangle \qquad (3.190)$$

with

$$v_j = \sum_{n=1}^{\infty} (-i)^n \int_{t_0}^{t} dt_1 \ldots \int_{t_0}^{t_{n-1}} dt_n QH(t_1) \ldots QH(t_n) a_j(t_n), \qquad (3.191)$$

where the optical potential v_j couples the P- and Q-spaces. This optical potential takes into account the flux of probability from the P-space to the Q-space. The zero-order approximation $v_j = 0$ reduces the system of coupled equations to the one corresponding to the SCE model. A local approximation has been used to evaluate v_j when the interaction time is small compared to a typical time of the system. In such a case it is assumed that two successive terms in the series (3.191) interact instantaneously, so that (3.191) gives

$$v_j = \sum_{n=1}^{\infty} (-i)^n \int_{t_0}^{t} dt' \, [QH(t')]^n \, a_j(t'). \qquad (3.192)$$

Target electron loss calculations have been obtained for impact of singly and multiply charged bare ions on hydrogen atoms, (*Ast* et al. (1988). At high enough collision energies these electron loss calculations give a good description of experimental total cross sections for ionization.

A two-center optical potential model has been also developed including target and projectile bound states in the P-space [*Lüdde* and *Dreizler* (1989)]. These states are selected to avoid overcompleteness and non-orthogonality of the sub-spaces subtended by the sets of target and projectile functions, which could result in double counting and incorrect interpretation of the population of different channels. In this model the population of ionization channels is determined by the loss of normalization of the P-space. Calculations [*Lüdde* and *Dreizler* (1989)] show good agreement with experimental total cross sections for impact of H^+ and He^{2+} ions on H targets at high enough collision energies. Different approximations to the exact optical potential has been recently considered [*Lüdde* et al. (1993), *Henne* et al. (1993)].

Singly- and doubly-differential cross sections for electron emission have been given using SCE close-coupling methods [*Schiwietz* (1990), *Martín* and *Salin* (1995)]. The use of the SCE approximations does not mean that two center effects in ionization are not accounted for. The use of an infinite complete basis of target states should permit covering the infinite space spanned by the solutions of the Schrödinger equation (3.182). Thus, the use of an appropriate and large basis of target states can give a good approximation to the description of two-center effects in target electron ionization.

Schiwietz (1990) proposed the description of the collision from a reference frame fixed on the target nucleus and using a linear combination:

$$\Psi_{tr}(\boldsymbol{x}, t) = \sum_{nlm} a_{nlm}(t) e^{-i\epsilon_{nl}t} \varphi_{nlm}(\boldsymbol{x}) + \sum_{lm} \int_0^\infty d\epsilon\, b_{lm}(\epsilon, t)\, e^{-i\epsilon t} \varphi_{\epsilon lm}(\boldsymbol{x}) \quad (3.193)$$

to represent the trial function. The eigenfunctions

$$\varphi_{nlm}(\boldsymbol{x}) = \frac{1}{x} u_{nl}(x)\, Y_{lm}(\hat{\boldsymbol{x}}) \quad (3.194)$$

satisfy the stationary Schrödinger equation

$$\left[\frac{1}{2} \frac{d^2}{dx^2} - V_T(x) - \frac{l\,(l+1)}{2x^2} + \epsilon_{nl} \right] u_{nl}(x) = 0 \quad (3.195)$$

as has been considered for the B1 approximation in Sect. 3.4.2. The subscript n is replaced by ϵ for continuum states. To describe high-energy continuum electrons, partial waves are summed up to $l = 8$. Moreover, the continuum is represented by a discrete sum of wave packets,

$$\psi_{lm}(\epsilon_1, \epsilon_2, x, t) = \int_{\epsilon_1}^{\epsilon_2} d\epsilon\, b_{lm}(\epsilon, t)\, e^{-i\epsilon t} u_{\epsilon l}(x) \quad (3.196)$$

so that

$$\sum_{lm} \int_0^\infty d\epsilon\, b_{lm}(\epsilon, t)\, e^{-i\epsilon t} \varphi_{\epsilon lm}(\boldsymbol{x})$$

$$\cong \sum_{jlm} \frac{1}{x}\, \psi_{lm}\!\left(\epsilon_j - \frac{\Delta\epsilon_j}{2}, \epsilon_j + \frac{\Delta\epsilon_j}{2}, x, t \right) Y_{lm}(\hat{\boldsymbol{x}}). \quad (3.197)$$

Assuming that the ionization is produced as a pulse at $t = 0$, the wave packet ψ_{lm} may be written as

$$\psi_{lm}(\epsilon_1, \epsilon_2, x, t) \cong \bar{b}_{lm}(\bar{\epsilon}, t) \int_{\epsilon_1}^{\epsilon_2} d\epsilon\, e^{-i\epsilon t} u_{\epsilon l}(x). \quad (3.198)$$

Using the fact that for $x\,\Delta\epsilon \ll \pi$ the radial wave function $u_{\epsilon l}$ may be considered independent of ϵ and for $\epsilon t \ll \pi$, in the exponential factor $e^{-i\epsilon t}$, ϵ may be replaced by its mean value $\bar{\epsilon}$ in the integral expression (3.198), and the wave packet ψ_{lm} can be further approximated as

$$\psi_{lm}(\epsilon_1, \epsilon_2, x, t) \cong \frac{2\bar{b}_{lm}(\bar{\epsilon}, t)}{t(\epsilon_2 - \epsilon_1)}\, e^{-i\bar{\epsilon}t} \sin\left(\frac{t(\epsilon_2 - \epsilon_1)}{2} \right) \int_{\epsilon_1}^{\epsilon_2} d\epsilon\, u_{\epsilon l}(x). \quad (3.199)$$

Contribution of high-lying Rydberg states are considered by renormalizing the energy width of the lowest-energy continuum states. Ionization probabilities are directly given by the square modules of the coefficient b_{lm} of the corresponding asymptotic target states. Details of the calculation procedure are given in the work of *Schiwietz* (1990), where singly- and doubly-differential cross sections for impact of protons on He targets are shown to give a good description of experimental data.

For this system, *Martín* and *Salin* (1995) have observed two-center effects in doubly differential cross sections. In their work they are mainly interested in the double excitation of He atoms, which manifests through a resonance effect in the spectrum of ejected electrons. Thus, they have chosen operators P and Q associated with the nonresonant and resonant (double excited) components of the He eigenstates, respectively. It is then assumed that the normalized eigenfunctions and eigenvalues of $P H_T P$ and $Q H_T Q$ are known for each value of the total angular momentum L and its z-component M.

In this model an initial ground state of the He target, with energy E_i, is considered. The final state with energy E corresponds to the case that one electron is in the ground $1s$ state of He^+ and the other electron is in continuum state with momentum $\boldsymbol{k}$. This wave function can be written as

$$\psi^- \left(\hat{\boldsymbol{k}}, E \right) = \sum_{LM} i^L e^{-i\delta_L} \left[Y_{LM} \left(\hat{\boldsymbol{k}} \right) \right]^* \psi_E^{LM}, \tag{3.200}$$

where δ_L is the phase shift [*Bachau* et al. (1991)]. The function ψ_E^{LM} is expanded as a linear combination of the resonant ϕ_j^{LM} and nonresonant χ_E^{LM} components, in the proximities of a resonance s, with angular momentum L [*Fano* (1961), *Feschbach* (1962)].

The transition amplitude can be written as

$$a_{if} (E) = \lim_{t \to \infty} \left\langle \psi^- \left(\hat{\boldsymbol{k}}, E \right) e^{-iEt} | \Psi_{tr} (t) \right\rangle . \tag{3.201}$$

The electron capture reactions by the projectile are neglected, so that the wave function $\Psi_{tr} (t)$ is chosen as a one-center expansion over eigenstates of $P H_T P$ and $Q H_T Q$,

$$\Psi_{tr} (t) = \sum_j a_j^{L_j M_j} (t) \, \phi_j^{L_j M_j} e^{-i\epsilon_j t} + \sum_{LM} \sum_n a_n^{LM} (t) \, \tilde{\chi}_n^{LM} e^{-iE_n t}. \tag{3.202}$$

The nonresonant wave functions χ_E^{LM} have been calculated in the static exchange approximation using the discretization of *Macías* et al. (1987, 1988) and *Bachau* et al. (1991), where $P H_T P$ is diagonalized in an L^2-integrable basis built from Slater-type orbitals. Thus, the resulting functions $\tilde{\chi}_n^{LM}$ are associated with a discrete spectrum and have the usual normalization for bound states. The lowest $\tilde{\chi}_n^{LM}$ functions represent He bound states. For $E > -2$ a.u., these functions are related to the exact continuum functions $\chi_{E_n}^{LM}$ for the same energy, in a finite domain of the configuration space,

$$\chi_{E_n}^{LM} = [\rho_L (E_n)]^{1/2} \, \tilde{\chi}_n^{LM} \tag{3.203}$$

with the state density $\rho_L (E_n)$ [*Macías* et al. (1988)].

The resonant wave functions $\phi_j^{L_j M_j}$ were calculated using the pseudo-potential Feschbach method [*Martin* et al. (1987a,b), *Bachau* et al. (1991)]

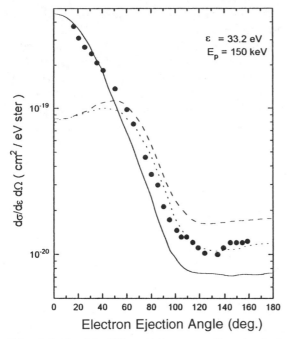

ε = 33.2 eV
E_p = 150 keV

Fig. 3.1. Double-differential cross sections for ionization of He by 150-keV protons as a function of the electron ejection angle. Solid circles: experiment, [*Bordenave-Montesquieu* et al. (1976)]. *Full curve*: close-coupling theoretical results, *short-dashed curve*: B1 results using correlated functions, *long-dashed curve*: B1 results with using Hartree-Fock-Slater initial and final wave functions. From *Martín* and *Salin* (1995)

with a Slater-type basis. Electron correlation is included in the resonant wave functions.

The method has been employed to determine double-differential cross sections for single ionization of He by impact of 150-keV protons [*Martín and Salin* (1995)]. The results are presented in Fig. 3.1 as a function of the electron ejection angle at a fixed final electron energy. They are in good agreement with experimental data [*Bordenave-Montesquieu* et al. (1976)]. Calculations are also compared with the B1 results obtained using the same basis as that for the close-coupling model or a Hartree-Fock-Slater description of the ejected electron. The close-coupling calculations show an increase (decrease) of the cross section at small (large) electron scattering angles with respect to the B1 predictions. This behavior is similar to the two-center results obtained with CDW-EIS calculations [*Stolterfoht* et al. (1987), *Fainstein* et al. (1988a)]. The important point is that this feature is obtained within a SCE model. It has been attributed to the fact that the projectile destroys the spherical symmetry of the target potential producing the coupling between continuum states with different values of L [*Martín and Salin* (1995)].

Finally, the forced impulse method (FIM) should be mentioned [*Reading and Ford* (1987)]. The method has been employed to study the impact of protons, antiprotons, and alpha particles on helium targets. At the beginning of the collision the system is considered to be in a correlated state. But in the following short time interval the system is described within the impulse approximation, where the influence of the residual target on the active electron is neglected. At the end of this interval the system collapses in a linear superposition of all possible correlated states. The same procedure is followed for the next time interval, and again a linear superposition is considered at the end of this second interval, and so on. In this way we define a crude mesh to describe the total collision time, forcing the validity of the impulse approximation in each one of the evolution steps. This method has been applied with success to describe two-electron emission at sufficiently high impact energies.

3.7 Dressed Projectiles

3.7.1 First-Order Theory by Bates and Griffing

The effect of the projectile electrons on the process of target ionization has been described within the framework of the first-order Born approximation by *Bates* and *Griffing* (1953), *Briggs* and *Taulbjerg* (1978) and *McGuire* et al. (1981). More recently, this subject has been revisited by *DuBois* and *Manson* (1990, 1993), *Manson* and *DuBois* (1992) and *McGuire* (1992) among other authors.

Let us consider the collision between a projectile dressed with N electrons and a monoelectronic target. Using the condition $M_P, M_T \gg 1$, the T-matrix element corresponding to the emission of a target electron is given by the expression:

$$T_{if}^{B1}(\boldsymbol{K}) = \frac{1}{(2\pi)^3} \int d^3R \, \exp\left(\mathrm{i}\boldsymbol{K} \cdot \boldsymbol{R}\right)$$

$$\times \left\langle \varphi_{Pf'}\left(\{\boldsymbol{s}_j\}\right) \varphi_{Tf}^-\left(\boldsymbol{x}\right) |V_{\mathrm{int}}| \varphi_{Pi'}\left(\{\boldsymbol{s}_j\}\right) \varphi_{Ti}\left(\boldsymbol{x}\right)\right\rangle \qquad (3.204)$$

with the perturbing potential:

$$V_{\mathrm{int}} = \frac{Z_t Z_p}{R} - \sum_{j=1}^{N} \frac{Z_t}{|\boldsymbol{R} + \boldsymbol{s}_j|} - \frac{Z_p}{s} + \sum_{j=1}^{N} \frac{1}{|\boldsymbol{R} + \boldsymbol{s}_j - \boldsymbol{x}|}, \qquad (3.205)$$

where $\{\boldsymbol{s}_j\}$ denotes the set of electron-projectile coordinates with respect to the projectile nucleus and φ_{Ti} and $\varphi_{Pi'}$ (φ_{Tf}^- and $\varphi_{Pf'}$) as the initial (final) target and projectile wave functions, respectively. In particular, φ_{Tf}^- is an incoming-Coulomb continuum wave function of the target.

Due to orthogonality of the target states the first two terms of V_{int} do not contribute to T_{if}^{B1}. Moreover, the third term of V_{int} does not contribute if a change of the projectile state occurs. Using a Fourier transformation similar to the one employed for bare projectiles, it can be shown that

$$T_{if}^{B1}(\boldsymbol{K}) = -\frac{1}{2\pi^2 K^2} \left\langle \varphi_{Pf'}(\{\boldsymbol{s}_j\}) \left| Z_p - \sum_{j=1}^{N} \exp(-i\boldsymbol{K}\cdot\boldsymbol{s}_j) \right| \varphi_{Pi'}(\{\boldsymbol{s}_j\}) \right\rangle$$
$$\times F_{0k}(-\boldsymbol{K}) \tag{3.206}$$

where $F_{0k}(-\boldsymbol{K})$ is the form factor for a final target continuum state of momentum $\boldsymbol{k}$. Then, the doubly-differential cross section for target ionization is given by

$$\frac{d\sigma}{d\epsilon d\Omega} = \frac{8\pi}{v^2} k \int\limits_{K_m}^{K_M} dK \frac{1}{K^3} Q_p^2 \, |F_{0k}(-\boldsymbol{K})|^2, \tag{3.207}$$

where

$$Q_p^2(\boldsymbol{K}) = \left| \left\langle \varphi_{Pf'}(\{\boldsymbol{s}_j\}) \left| Z_p - \sum_{j=1}^{N} \exp(-i\boldsymbol{K}\cdot\boldsymbol{s}_j) \right| \varphi_{Pi'}(\{\boldsymbol{s}_j\}) \right\rangle \right|^2 \tag{3.208}$$

is an effective charge depending on the momentum transfer. For bare projectiles it follows that $Q_p^2 = Z_p^2$. It should be recalled that the case where the projectile state remains in its initial state during the collision is referred to the monoelectronic process because only the target electron is removed,

$$Q_p^2(\boldsymbol{K}) = \left| Z_p - \left\langle \varphi_{Pi'}(\{\boldsymbol{s}_j\}) \left| \sum_{j=1}^{N} \exp(-i\boldsymbol{K}\cdot\boldsymbol{s}_j) \right| \varphi_{Pi'}(\{\boldsymbol{s}_j\}) \right\rangle \right|^2. \tag{3.209}$$

In particular, for $K \to 0$ (separate encounters), we obtain $Q_p^2 = (Z_p - N)^2$. In this case the projectile electrons screen the corresponding nuclear charge.

The case where the projectile state changes is referred to as the dielectronic process,

$$Q_p^2(\boldsymbol{K}) = \left| \left\langle \varphi_{Pf'}(\{\boldsymbol{s}_j\}) \left| \sum_{j=1}^{N} \exp(-i\boldsymbol{K}\cdot\boldsymbol{s}_j) \right| \varphi_{Pi'}(\{\boldsymbol{s}_j\}) \right\rangle \right|^2, \quad i \neq f \tag{3.210}$$

where only the interaction between the projectile electrons and the target electron contributes to the target electron ionization. It is evident that as $K \to 0$ this term does not give any contribution. If all final projectile states are summed to calculate Q_p^2 [$McGuire$ et al. (1981)], we obtain $Q_p^2 \to Z_p^2 + N$ for $K \to \infty$ (close encounters). Hence, the projectile nucleus and electron act incoherently, so that the target electron feels separately the influence of these particles. It should be recalled that this effect originates from the interaction between the projectile electrons and the target electron.

For the particular case of a projectile dressed with only one electron, the double-differential cross sections corresponding to monoelectronic and dielectronic processes can be expressed in the simple form [*DuBois* and *Manson* (1990), *Manson* and *DuBois* (1992)]

$$\left(\frac{d\sigma}{d\epsilon d\Omega}\right)_M = \frac{8\pi}{v^2} k \int_{K_m}^{K_M} dK \frac{1}{K^3} \left|Z_p - F_{i'i'}(\boldsymbol{K})\right|^2 |F_{if}(-\boldsymbol{K})|^2 , \quad (3.211)$$

$$\left(\frac{d\sigma}{d\epsilon d\Omega}\right)_D = \frac{8\pi}{v^2} k \int_{K_m}^{K_M} dK \frac{1}{K^3} \left[1 - |F_{i'i'}(\boldsymbol{K})|^2\right] |F_{if}(-\boldsymbol{K})|^2, \quad (3.212)$$

where primed and unprimed quantities refer to the projectile and target, respectively, and the definition of the form factor has been extended for any final bound or continuum state,

$$F_{if}(\boldsymbol{q}) = \int d^3r \, \varphi_f^*(\boldsymbol{r}) \exp(-i\boldsymbol{q} \cdot \boldsymbol{r}) \, \varphi_i(\boldsymbol{r}). \quad (3.213)$$

It is obvious that for a final continuum target state with momentum $\boldsymbol{k}$, $F_{if}(\boldsymbol{q}) = F_{ok}(\boldsymbol{q})$.

If more than one projectile electron are present, their contributions are to be included in the cross section formulae (3.211,212). In the monoelectronic process $F_{i'i'}$ is replaced by a sum of form factors:

$$F_{i'i'} \rightarrow \sum_s N_p^{(s)} F_{i'i'}^{(s)}, \quad (3.214)$$

where $F_{i'i'}^{(s)}$ is associated with an electron in the sub-shell s of the incident ion and $N_p^{(s)}$ is the corresponding occupation number of that sub-shell. On the other hand, for the dielectronic process, $(d\sigma/d\epsilon\, d\Omega)_D$ is replaced by a sum of cross sections [*DuBois* and *Manson* (1990)]:

$$\left(\frac{d\sigma}{d\epsilon d\Omega}\right)_D \rightarrow \sum_s N_p^{(s)} \left(\frac{d\sigma}{d\epsilon d\Omega}\right)_D^{(s)}, \quad (3.215)$$

where $(d\sigma/d\epsilon\, d\Omega)_D^{(s)}$ is associated with an electron in the sub-shell s.

The influence of the target electron on the ionization of the projectile electron can be obtained by exchanging the quantities associated with the projectile and target. A complete review on the role of two-center electron-electron interaction in projectile electron loss has recently been given by *Montenegro* et al. (1994).

3.7.2 Higher-Order Theories for Binary Encounter Electrons

The prominent feature of the spectrum of fast electrons ejected in the forward direction is the binary-encounter peak. For bare projectiles the BE peak is described by the elastic scattering model in conjunction with the Rutherford cross section involving a Z_p^2 scaling, [*Lee* et al. (1990)]. However, *Richard* et al. (1990) observed for the forward electron ejection that the BE peak increases as the projectile charge q decreases. Different explanations are possible for the enhancement of the binary encounter peak. One interpretation is that the outer shell electrons of dressed projectiles reduce the incident target electron velocity, increasing thus the Rutherford cross section and, hence, the corresponding BE peak intensity.

On the contrary, inner electrons screen the projectile nuclear charge, in this way decreaing the BE cross section [*Reinhold* et al. (1990a), *Taulbjerg* (1990b), *Ponce* et al. (1993)]. In the cases of F^{q+} impact on H_2 and He, measured by *Richard* et al. (1990), the effect of the outer projectile electrons dominate the contribution of the inner ones. However, for different systems the dominance may be reversed. CTMC calculations for the $U^{q+} + Ar$ system show that inner electrons can dominate the outer ones for small enough values of q [*Olson* et al. (1990)]. In the CTMC calculations, model potentials [*Green* et al. (1969), *Garvey* et al. (1975)] have been used to describe the interactions of the active target electron with the residual target core and with the dressed projectile.

The exchange interaction between target and projectile electrons has also been considered in different theoretical analysis [*Taulbjerg* (1990b, 1991), *Chen* et al. (1991), *Bhalla* and *Shingal* (1991), *Ponce* et al. (1993), *Macek* and *Taulbjerg* (1993)]. Thus, exchange contributions, which correspond to the capture of a target electron and the simultaneous emission of a projectile electron in the BE peak region, have been determined.

We should also mention that the BE approximation [*Bonsen* and *Vriens* (1970)] which differs slightly from the elastic scattering model developed by *Lee* et al. (1990) in the definition of the final electron energy, has been employed with success by *Reinhold* et al. (1990b) to describe the enhancement of the BE peak observed experimentally [*Richard* et al. (1990)].

In the following we develop the model introduced by *Ponce* et al. (1993) from which other theoretical approximations can be obtained. For simplicity the analysis is reduced to the impact of a monoelectronic projectile on a light target. Hence, the BE electron production is simulated by the elastic scattering of a free target electron by a hydrogenic projectile. The characteristic feature of the different theoretical approaches is that the electron-ion interaction is treated in higher-order.

We describe the process from the projectile nucleus and denote with s_j the position vector of the electron j with respect to the projectile nucleus. The Schrödinger equation can be written as:

$$\left[H(s_1) + H(s_2) + \frac{1}{s_{12}} - E \right] \Psi(s_1, s_2) = 0, \qquad (3.216)$$

where s_{12} is the position vector of electron 1 with respect to electron 2, and

$$H(s) = -\frac{1}{2}\nabla_s^2 - \frac{Z_p}{s}. \tag{3.217}$$

The initial projectile ground state satisfies the equation:

$$(H(s) - \epsilon_i)\,\varphi_i(s) = 0, \tag{3.218}$$

where $E = \epsilon_i + p^2/2$, with ϵ_i and p being the initial binding energy of the projectile electron and momentum of the target electron with respect to the projectile, respectively. We restrict our analysis to electron energies that verify the condition $p^2 < 0.75\,Z_p^2$, so that projectile excitation is neglected. Considering the possibility of electron exchange, only two channels are open. Then, the wave function describing the process is given by the linear combination:

$$\Psi^\pm(s_1, s_2) = 2^{-1/2}\left[F^\pm(s_1)\,\varphi_i(s_2) \pm F^\pm(s_2)\,\varphi_i(s_1)\right], \tag{3.219}$$

where the superscripts (+) and (-) correspond to singlet and triplet initial spin states.

The amplitudes $F^\pm(s)$ must satisfy the asymptotic limit:

$$\lim_{s\to\infty} F^\pm(s) \approx (2\pi)^{-3/2}\,\{\exp[i\boldsymbol{p}_i\cdot\boldsymbol{s} + i\eta\ln(ps - \boldsymbol{p}_i\cdot\boldsymbol{s})]$$

$$+ \frac{f^\pm(p,\theta)}{s}\,\exp[ips - i\eta\ln(2ps)]\}, \tag{3.220}$$

where we have used the elastic condition $p = |\boldsymbol{p}_i| = |\boldsymbol{p}_f|$ and the definitions $\eta = -(Z_p - 1)/p$ and $\cos\theta = \hat{\boldsymbol{p}}_i\cdot\hat{\boldsymbol{s}}$. Replacing (3.219) in (3.216), the amplitudes $F^\pm(s)$ can be obtained by solving the equation:

$$\left(-\frac{1}{2}\nabla_s^2 - \frac{Z_p}{s} + V(s) - \frac{1}{2}p^2\right)F^\pm(s)$$

$$= \mp\left\langle\varphi_i(s\prime)\left|\frac{1}{|\boldsymbol{s} - \boldsymbol{s}'|} + \epsilon_i - \frac{1}{2}p^2\right|F^\pm(s')\right\rangle\varphi_i(s) \tag{3.221}$$

with

$$V(s) = \left\langle\varphi_i(s')\left|\frac{1}{|\boldsymbol{s} - \boldsymbol{s}'|}\right|\varphi_i(s')\right\rangle. \tag{3.222}$$

If in (3.221), we consider

$$-\frac{Z_p}{s} + V(s) = -\frac{(Z_p - 1)}{s} + w(s) \tag{3.223}$$

with $w(s)$ the short range potential

$$w(s) = -\left(\frac{1}{s} + Z_p\right)e^{-2Z_p s} \tag{3.224}$$

and we take $w(s)$ and the right-hand side of (3.221) as perturbations [*Ponce* et al. (1993)], the scattering amplitudes defined in (3.220) can be approximated in a distorted wave approximation as

$$f^{\pm}(p,\theta) = f^c_{Z_p-1}(p,\theta) + f^d(p,\theta) \pm f^{ex}(p,\theta). \qquad (3.225)$$

Here, $f^c_{Z_p-1}$ is the Coulomb scattering amplitude corresponding to the dispersion of the target electron in the projectile field of a net charge $q = Z_p - 1$. f^d is the direct amplitude which takes into account the dispersion due to the short range potential $\omega(s)$,

$$f^d(p,\theta) = -(2\pi)^2 \left\langle \psi^-_{p_f}(s) \, |\omega(s)| \, \psi^+_{p_i}(s) \right\rangle, \qquad (3.226)$$

where $\psi^{\pm}_p$ are outgoing and incoming Coulomb waves in a potential of charge $-(Z_p - 1)$, and f^{ex} is the exchange amplitude

$$f^{ex}(p,\theta) = f^{ex}_{12}(p,\theta) + f^{ex}_2(p,\theta) \qquad (3.227)$$

with

$$f^{ex}_{12}(p,\theta) = -(2\pi)^2 \left\langle \psi^-_{p_f}(s_1)\,\varphi_i(s_2) \left| \frac{1}{s_{12}} \right| \psi^+_{p_i}(s_2)\,\varphi_i(s_1) \right\rangle, \qquad (3.228)$$

and

$$f^{ex}_2(p,\theta) = +(2\pi)^2 \left\langle \psi^-_{p_f}(s_1)\,\varphi_i(s_2) \left| \frac{1}{s_2} \right| \psi^+_{p_i}(s_2)\,\varphi_i(s_1) \right\rangle. \qquad (3.229)$$

The scattering amplitude $f^c_{Z_p-1}$ is obtained as

$$f^c_{Z_p-1}(p,\theta) = -\eta\, e^{2i\sigma_0} \frac{\exp\left[-i\eta \ln\left(\sin^2(\theta/2)\right)\right]}{2p \sin^2(\theta/2)}, \qquad (3.230)$$

where $\sigma_l = \arg \Gamma(l + 1 + i\eta)$ is the Coulomb phase shift [*McDowell and Coleman* (1970)]. The differential cross section corresponding to the elastic scattering of the electron in a Coulomb field of charge $(Z_p - 1)$ is given by the Rutherford formula (3.2)

$$\frac{d\sigma}{d\Omega} = \left|f^c_{Z_p-1}\right|^2 = \frac{(Z_p - 1)^2}{4p^4 \sin^4(\theta/2)}. \qquad (3.231)$$

The evaluation of the amplitude f^d can be done using a partial-wave analysis, so that [*Taulbjerg* (1990b)]

$$f^d(p,\theta) = \frac{1}{p} \sum_l (2l + 1)\,(\exp(2i\hat{\delta}_l) - 1)\,\exp(2i\sigma_l)\, P_l(\cos\theta), \qquad (3.232)$$

where $\hat{\delta}_l$ is the phase shifts of the short range potential,

$$\hat{\delta}_l = \frac{2}{p} \int_{l_0}^{\infty} ds \left(\frac{1}{s} + Z_p\right) \exp(-2Z_p s)\,(F_l(p,s))^2. \qquad (3.233)$$

Here, $F_l(p,s)$ is the radial Coulomb wavefunction normalized to unit amplitude at large distances. The differential cross section for the scattering of the electron in the screening field given by (3.223) results from the expression

$$\frac{d\sigma}{d\Omega} = \left| f^c_{Z_p - 1} + f^d \right|^2.$$
(3.234)

The terms f^{ex}_{12} and f^{ex}_2 can be also obtained using partial-wave analysis as it has been done by *Taulbjerg* (1990b) and *Ponce* et al. (1993). The complete differential cross section is given by

$$\frac{d\sigma}{d\Omega} = \frac{1}{4} \left| f^+ \right|^2 + \frac{3}{4} \left| f^- \right|^2.$$
(3.235)

For multielectron projectiles, the importance of the inclusion of the term f^d in the calculation of BE cross sections was determined by *Shingal* et al. (1990), neglecting the exchange mechanisms. The role played by exchange contributions was analyzed by *Taulbjerg* (1990b) as well as by *Bhalla* and *Shingal* (1991). *Taulbjerg* (1990b) has neglected the exchange contribution f^{ex}_2 considering that this term cancels if the states φ_i and $\psi^\pm_p$ are orthogonal. However, this condition can only be accomplished if the interaction $1/s_{12}$ is neglected, so that the term f^{ex}_2 should also be turned off. The inclusion of the term f^{ex}_2 was shown to improve the agreement between theoretical results and experimental cross sections [*Ponce* et al. (1993)]. A more elaborate calculation was performed by *Bhalla* and *Shingal* (1991), who have solved (3.221) using a Hartree-Fock self-consistent field approximation.

Taulbjerg (1991) has described the reaction for a monoelectronic projectile using a B1-PW model, where the final state of the ionized electron is represented by a plane wave (Sect. 3.4.1). Moreover, *Macek* and *Taulbjerg* (1993) have made a theoretical analysis of the DSPB approximation for the case of monoelectronic projectiles. They have considered an initial wave function:

$$\Phi^{DSPB}_i(s_1, s_2) = (2\pi)^{-3/2} \int dk_i \tilde{\varphi}_T(k_i) \exp[i(\mu_T K_i + k_i) \cdot R_f] \gamma^+(k_i)$$
$$\times \Psi^+[(Z_P - 1), \mu_P(k_i - v), s_1] \varphi_P(s_2),$$
(3.236)

where we have used the same notation as in Sect. 3.5.4, and a final wave function

$$\Phi^{DSPB}_f(s_j, s_r) = \Phi^{OBK}_f(s_j) \varphi_P(s_r); \quad j, r = 1, 2, \quad j \neq r$$
(3.237)

where Φ^{OBK}_f is defined in Sect. 3.4.3. The quantity x_j (s_j) denotes the position vector of the j-electron with respect to the target (projectile) nucleus, and φ_T (φ_P) stands for an electron bound state of the target (projectile). Furthermore, they have considered direct and exchange matrix elements given by

$$T^d_{if} = \left\langle \Phi^{DSPB}_f(s_1, s_2) \left| -\frac{Z_t}{x_1} + \frac{1}{s_{12}} - \frac{1}{s_1} \right| \Phi^{DSPB}_i(s_1, s_2) \right\rangle$$
(3.238)

and

$$T^{ex}_{if} = \left\langle \Phi^{DSPB}_f(s_2, s_1) \left| \frac{1}{s_{12}} \right| \Phi^{DSPB}_i(s_1, s_2) \right\rangle,$$
(3.239)

where contributions due to the nonorthogonality of continuum and bound states of the projectile electron are neglected. The term associated with the perturbing potential $-Z_t/x_1$ in (237) may be approximated [*Macek and Taulbjerg* (1993)] as

$$T_{if}^\circ \cong T^R(\boldsymbol{p}, \boldsymbol{p} - \boldsymbol{K})\, \gamma^+(k_i)\, \tilde{\varphi}_T(\boldsymbol{k}_i) \qquad (3.240)$$

with $\boldsymbol{k}_i$ the initial momentum of the target electron and $T^R(\boldsymbol{p}, \boldsymbol{q})$ the off-energy-shell T-matrix element for the Rutherford scattering of the electron in the projectile field, given by

$$T^R(\boldsymbol{p}, \boldsymbol{q}) = -\frac{1}{2\pi^2}(Z_p - 1)\frac{1}{|\boldsymbol{p} - \boldsymbol{q}|^2}\left(\frac{|\boldsymbol{p} - \boldsymbol{q}|^2}{(p+q)^2}\right)^{i\varsigma}\frac{\Gamma(1 - i\zeta)}{\Gamma(1 + i\zeta)} \qquad (3.241)$$

with $\zeta = (Z_p - 1)/p$. The terms in (3.238) containing the perturbation $(1/s_{12} - 1/s_1)$ represent screening corrections to the transition matrix element. Also, the exchange matrix element can be written as

$$T_{if}^{ex} = -(2\pi)^{-2}f_{12}^{ex}\, \gamma^+(k_i)\, \tilde{\varphi}_T(\boldsymbol{k}_i), \qquad (3.242)$$

where f_{12}^{ex} is defined as in (3.228). In the BE peak, the on-energy-shell limit of (3.242) must be considered, so that $\gamma^+(k_i) \cong 1$ and then the matrix elements T_{if}^0 and T_{if}^{ex} are immediately associated as direct and exchange terms of the intuitive elastic scattering model.

4. Experimental Methods

The experimental techniques used to study differential electron emission by heavy particle impact are similar to those used for proton impact. Since the proton impact studies have recently been reviewed by *Rudd* et al. (1992), the reader is referred to that article for details concerning experimental techniques and difficulties. Here we provide a general overview supplemented with information pertinent to dressed ion impact and details not discussed in the previous review.

4.1 Experimental Set-Up

Briefly, the experimental technique is to accelerate a heavy particle beam to the desired energy, magnetically select one of the charge state components, and inject the collimated beam component into a target chamber where it passes through a target region and is collected in a Faraday cup. The number of impacting particles is determined from the integrated charge and the known charge state of the projectile ions. In order to measure the emission of lower energy electrons, the magnetic field within the chamber must be reduced to a few milligauss or less. This general concept is shown schematically in Fig. 4.1.

Electrons produced via interactions with a target gas, exit the interaction region at specific angles with respect to the beam direction. An electron spectrometer is used to measure their energy and they are individually detected by a particle counter. Generally, electrostatic energy analyzers, positioned to observe a specific angle (or angles) of emission are used and the analyzing voltage is scanned in order to measure the yield of electrons per unit target density and incoming projectile. Absolute doubly-differential ionization cross sections are determined from these yields either (a) by directly measuring the electron detection efficiencies, solid angles, and other experimental parameters, or (b) by indirectly obtaining these parameters via normalization using other collision systems and/or impact energies where absolute cross sections are known. The majority of studies have placed their data on an absolute scale via normalization to known cross sections in order to circumvent problems associated with determining absolute target densities and electron detection efficiencies.

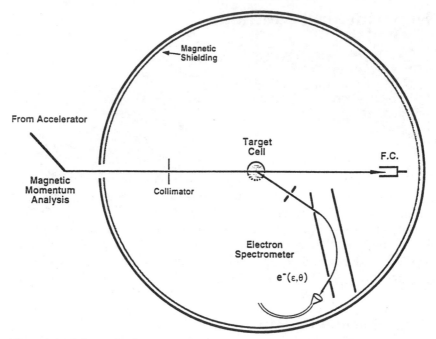

Fig. 4.1. Schematic diagram of a typical electron emission experiment employing a differentially pumped target cell and a parallel-plate electrostatic electron spectrometer which can be rotated about the scattering center

Overall uncertainties in the absolute cross sections are due to uncertainties in the target density, detection efficiency, etc. In regions where the measured signals are large, the overall uncertainties are typically on the order of ±30 %. At the extremes of the reported electron energy ranges, considerably larger uncertainties are quoted or should be expected. For example, at the highest emission energies the ionization cross sections become small. Thus, statistical uncertainties become large and scattered background contributions can influence the accuracy of the data. For low emission energies, approximately 10 eV and lower, large uncertainties are again expected because of experimental difficulties that will be discussed later in this section.

We now provide specific details pertaining to gaseous target design, techniques useful for studying the emission of low-energy electrons, and coincidence methods which are applicable to studying the various ionization mechanisms in heavy particle-atom collisions.

4.2 Target Design

Two types of gaseous targets, shown schematically in Fig. 4.2, have been used to study electron emission induced by heavy particle impact. In the figure,

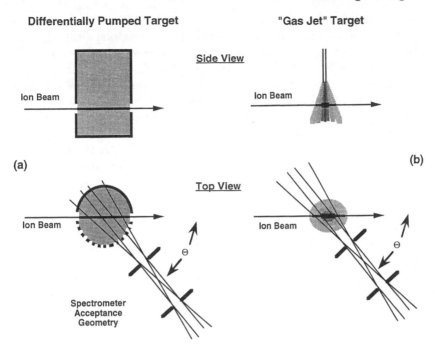

Fig. 4.2. Schematic views of a differentially pumped (a) and a directed gas (b) target. The thickness of the ion beam lines indicates the relative electron production probabilities. Target gas densities are indicated by various degrees of shading. Spectrometer acceptance geometries are indicated in the overhead views

relative electron emission intensities are indicated by the thickness of the "ion beam line." These intensities are proportional to the target density, which is indicated by the degree of shading. The specific type of target chosen by different groups depends on many criteria. For example, important factors are (i) the target and background density ratios that are required, (ii) whether a point- or an extended-source target is desired, and (iii) whether independent measurements of absolute cross sections are intended.

One type of target, as illustrated in Fig. 4.2a, confines the target gas in a differentially pumped cell which has either a continuous slit or fixed apertures in order for the electrons to exit. The major advantage of this type of target is the ability to determine absolute target densities, which is essential for independently determining absolute cross sections. Another advantage is that containment of the target gas within the target region reduces the number of electrons produced outside the target cell and, hence, reduces background signals.

A disadvantage of a differentially pumped target is the large surface areas of the target cell and its mounting brackets in the vicinity of the interaction region. These surfaces and the materials used can produce perturbations in the electron emission due to surface charging, residual magnetic field effects, or because electrons scatter from the surfaces. Other possible problems are that large background signals can be produced if the beam does not pass completely uninterrupted through the entrance and exit apertures of the cell, and that the length of the interaction region viewed by the spectrometer is a function of the angle of observation (Fig. 4.2). In certain experiments, another concern is that an extended, rather than a point, source of electrons is produced. Also of significance for lower energy electron emission, is that the electrons must travel for some distance through a fairly dense target gas before they exit the cell into a region of higher vacuum. This introduces a finite probability that the electrons can be scattered out of, or other electrons can be scattered into, the proper trajectory for detection. To quote from *Rudd* et al. (1992), "One detector's uncounted, absorbed electron is another detector's background."

Several of these disadvantages can be overcome by using a directed-gas target, shown schematically in Fig. 4.2b. Using a directed-gas target minimizes problems associated with large surface areas in the vicinity of the interaction region. Also, in principle, a point source of electrons can be achieved, which is a desirable feature in many experiments. A point source also minimizes electron absorption effects, as discussed above. However, as we will demonstrate, a point source is difficult to achieve.

Unfortunately, directed gas targets suffer from several major disadvantages. First, it is not possible to directly determine the absolute target density; this requires using a normalization procedure to place the cross sections on an absolute scale. Second, the target density decreases rapidly with distance, both vertical and horizontal, from the gas source and does not suddenly drop to zero at the edge of the "jet". This means that alignment with respect to the beam is critical and that the spectrometer can view different target thicknesses at different angles of emission.

We mentioned above that a point target is difficult to achieve and directed-gas targets require careful alignment with respect to the beam. As pointed out in *Rudd* and *Macek* (1972) and *Rudd* et al. (1992), the measured electron signal is proportional to the beam intensity, the target density, and the solid angle for detection. For finite beams and targets, the product of these quantities must be determined at each point of electron production and then summed. Thus one obtains the effective term $(nl)_{eff}$ discussed in detail by *Rudd* et al. (1992),

$$(n\,l\,\Omega)_{eff} = \sum_{x,y,z} N_B(x,y,z) N_T(x,y,z) \Delta\Omega(x,y,z), \qquad (4.1)$$

where n is the target density, N_B, N_T are the beam and target densities, and $\Delta\Omega$ is the solid angle for detection of a particle originating from a particular source point. Note that the summation is over all values of x, y, z.

For illustration purposes, let us assume that the beam is an infinitely fine line of constant intensity along z and is located at x_o, y_o. Also, assume that the gas source is located in the x, y plane at z_o. In this case (4.1) becomes:

$$(n\,l\,\Omega)_{eff} \approx N_B \sum_{x,y,z} N_T(x,y,z)\Delta\Omega(x,y,z). \qquad (4.2)$$

The summation over z must still be performed because of the dependence on z in the N_T and $\Delta\Omega$ terms. This summation is critical when measuring electron emission at $0°$ and $180°$ where the spectrometer views the entire length of the beam. Since N_T generally does not immediately drop to zero for $z \neq z_o$, the summation must be taken along the entire beam path (at least along the beam path until a physical aperture is encountered, for only then does $\Delta\Omega \to 0$). In this summation, N_T must include both the target and residual gas contributions.

The summation is also important when a point-source target is required. For example, a directed-gas target having a roughly constant pressure of 1 mTorr inside a 1 mm diameter cylinder and linearly decreasing to the background pressure of 0.05 mTorr at a distance of 1 mm outside this cylinder might be considered to be a good approximation to a point source at z_0. However, if the final collimating aperture for the beam is located 10 cm upstream from the target and the entrance aperture of the spectrometer is 10 cm downstream and we assume that the spectrometer is equally sensitive to all solid angles, 90 % of the electron signal will come from background contributions, i.e., from electrons generated in interactions at $z \neq z_0$ along the beam path. We stress that this example is overly simplified. In particular, our assumption of a constant $\Delta\Omega$ along the beam path is incorrect. For example, in certain spectrometers $\Delta\Omega$ is largest directly at the spectrometer entrance; which can make background contributions at this point much more important than electron production from the more distant directed-gas target.

Equation (4.1) also indicates why directed-gas targets must be designed to provide highly reproducible target beam conditions. For most of the directed source targets that have been used, the target gas significantly expands and the target density, N_T, rapidly decreases in the x, y, z directions. Thus any change in the horizontal or vertical positions of the beam strongly influences the measured electron signal. For these reasons, it is essential that the parameters in (4.1) are known and that the conversion from measured electron signal to cross section incorporates the summation shown.

To summarize: (i) Higher target and/or background density ratios can be achieved with a differentially-pumped target. (ii) If a point-source is required, it can only be approximated by a well-designed directed-gas target. (iii) Direct, independent measurements of absolute cross sections require the knowledge of the target density. This requires well-designed differentially-pumped targets. (iv) A directed-gas target is more applicable to the measurement of low-energy electrons since there are fewer surfaces in the source region which can lead to problems. (v) The ease in constructing and using either type of target is mainly due to personal preferences. Careful alignment is required in both cases, but for different reasons. To conclude, both differentially-pumped and directed-gas targets have their advocates and, with proper care, data produced by both techniques are of comparable quality.

4.3 Electron Energy Analysis

To date, most studies of differential electron emission induced by heavy particle impact have employed electrostatic electron energy analysis. Exceptions are studies of extremely high-energy electron emission in energetic heavy particle collisions where magnetic energy analysis has been used [*Güttner* et al. (1982)]. But information obtained in these high energy studies is outside the scope of this review and will not be discussed. In Fig. 4.1 an example of an electron emission apparatus which uses a single electrostatic spectrometer that can be rotated around the scattering center in order to investigate different angles of emission is illustrated. Improved efficiency in the collection of data and minimization of several experimental uncertainties can be achieved by designing the spectrometer to measure several electron emission angles simultaneously, [*Varga* et al. (1992)], as illustrated in Fig. 4.3.

Electrostatic energy analyzers are used since they are amenable for measuring a broad range of electron energies. Generally they have an poorly-defined lower energy limit of approximately 10 eV. This lower limit is due to problems associated with transmitting and detecting low-energy electrons and because the background signals increase rapidly at low energies. An equally

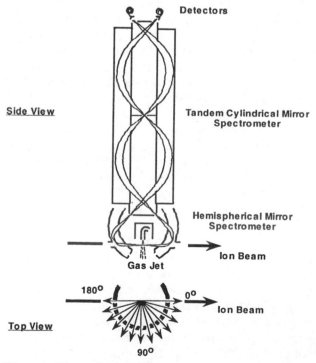

Fig. 4.3. The multiangle electron spectrometer developed at ATOMKI. From *Varga* et al. (1992)

ill-defined upper energy limit arises from the inability to apply higher and higher voltages to the spectrometer without causing electrical breakdown. Typically this upper limit has been 5–10 keV.

The upper limit generally has not been overly restrictive since it permits probing electron emission for the smallest impact parameters in collisions up to a few MeV/u. The lower limit, however, is more problematic because low-energy electron emission dominates both the target and the projectile ionization cross sections in their respective reference frames. Thus, difficulties associated with measuring low-energy electrons means that the vast majority of target ionization interactions have rarely been investigated. For target ionization, this uninvestigated region corresponds to electrons emitted in distant collisions. On the other hand, distant collisions are routinely investigated for projectile ionization because low-energy electrons emitted in the projectile frame of reference are kinematically shifted to higher energies in the laboratory frame and measuring higher energy electrons is easier. In contrast, for projectile ionization it is the higher energy electrons, resulting from close collisions, that are difficult to distinguish. This is because they merge with the "background" resulting from target ionization.

To reduce the low-energy limit, experimental difficulties associated with (a) the rapid decrease in the real ejected electron signal, and (b) the increase in the background signal as the electron energy approaches zero, need to be overcome. The decrease in the real signal arises from a decrease in the efficiency of transporting electrons from the interaction region to the detector due to the influence of residual magnetic and/or electrostatic fields. These factors normally overshadow any decrease in the detector response to low-energy electrons since the detector is generally biased to pre-accelerate the electrons just prior to their detection. The increased background signals at low energies result from higher energy electrons which are scattered and/or energy degraded in such a way that they eventually reach the detector.

4.3.1 Low-Energy Electron Emission

Since low-energy electron emission dominates the total ionization cross sections and because this region has been investigated in relatively few cases, we will discuss some of the methods that can be (or have been) used.

By far the best method for accurately measuring low-energy electron emission is to use time-of-flight energy analysis. This method is extremely effective because, with respect to electrostatic energy analyzers, time-of-flight spectrometers typically reduce both the distance the electrons must travel and the surfaces with which they come into contact. Thus, the influence of residual electrostatic and magnetic fields on the electron trajectories is considerably reduced. The disadvantage of using time-of-flight energy analysis is that achievable time resolutions provide poor energy resolution above 100 eV-unless a long flight path is used which then reduces this method's usefulness for studying low-energy electron emission. In spite of the obvious advantages, until recently no one had applied this method to ionization induced by heavy ions, the only ion impact data being for proton impact [*Toburen* and *Wilson* (1975), *Skogvall* and *Schiwietz* (1990)]. However, recently the recoil ion spectroscopy method has been extended to include measurements of low-energy electrons [*Moshammer* et al. (1994, 1996)].

An alternative method, compatible with electrostatic energy analysis, is to accelerate the electrons to higher energies as they exit the target region. This improves the transmission of low-energy electrons from the interaction region to the detector. *DuBois* (1993, 1994) and *DuBois* and *Manson* (1994) used this approach to study low-energy emission for hydrogen, helium, and carbon particle impact on helium. A specially designed differentially-pumped target cell, consisting of an inner, negatively biased, cylinder surrounded by an outer, grounded, shield slightly larger in diameter, was used. High transmission grids, placed over the slits of both cylinders, produced a uniform field for accelerating electrons to higher energies as they exited the target region. Considerable improvement in the detection of low-energy electrons was found as the bias voltage was increased from zero to a couple volts. Using a bias of -2 volts, cross sections down to 0.5 eV electron emission energies were reported. However, when using this technique one should be careful to avoid acceleration fields which produce large changes in the electron velocities, as this can introduce artifacts in the electron emission spectra due to defocusing effects.

4.3.2 Background Suppression at Low Energies

As stated above, measurements of low-energy electron emission are hampered by scattered electron background signals. For example, electrons scattering from gases and surfaces within the target region or vacuum chamber can eventually enter the spectrometer and be detected, unless adequate shielding and collimation are used. For these reasons extreme care must be taken whenever low-energy electron emission measurements are attempted.

Suárez et al. (1993a,b) investigated the low energy electron emission from Ne. They used a directed-gas target and performed measurements as a function of target source distance above the ion beam. In the 0–10 eV range, they found that the normalized count rate systematically increased as the source was moved farther away from the ion beam. At the time, this effect was tentatively interpreted as arising from spurious signals associated with electrons scattering in an extended gas target. Specifically, the authors suggested that electrons originally emitted in other directions undergo collisions in the extended target and are scattered into the electron spectrometer where they contribute to the measured signal. This is the electron scattering that was discussed previously. By placing the directed-gas target extremely close to the ion beam, a better approximation to a point source is achieved and the false signals are minimized. This interpretation has been re-analyzed [*Bernardi* et al. (1996)] and it appears that an explanation different from rescattering of the electrons is necessary to explain the observations. We would expect that the "extended gas target effect" observed by *Suárez* et al. (1993a,b) is specific for their apparatus. Their data do, however, illustrate that care must be taken in low-energy electron emission measurements and that it cannot be taken for granted that a directed-gas target is a point source where electron absorption can be ignored.

Spurious background signals in the low-energy region can also be produced by electrons scattering from surfaces within the spectrometer [*Bernardi* and *Meckbach* (1995), *Irby* (1995), and *Bernardi* et al. (1996)]. This problem is particularly important when a parallel plate spectrometer is used to investigate low-energy electron emission. In this case, a small electrostatic field is applied to the spectrometer which means that the trajectories of the vast majority of higher-energy electrons will be relatively unaffected. Thus higher energy electrons will strike the spectrometer back plate and the field straightener plates. Also, low-energy electrons will strike the front plate. In all cases, these electrons can scatter from the surfaces or they can produce secondary electrons. The electric field used for energy analysis will sweep the scattered and secondary electrons toward the front plate, i.e., toward the exit slit of the spectrometer. This can introduce an additional background signal, particularly in the low-energy electron region.

To demonstrate this we perform an estimation of the number of backscattered electrons. The present analysis is similar to that by *Bernardi* and *Meckbach* (1995). The real electron signal, $N_s(\epsilon_s, \theta)$, observed at an energy ϵ_s and angle θ, is given by:

$$N_s(\epsilon_s, \theta) = c_N \frac{d^2\sigma(\epsilon_s, \theta)}{d\epsilon d\Omega} \Delta\Omega_{out} \Delta\epsilon, \qquad (4.3)$$

where c_N is a proportionality factor, $d^2\sigma/d\epsilon\, d\Omega$ is the doubly-differential cross section for electron emission, $\Delta\Omega_{out}$ is the solid angle for detection and $\Delta\epsilon$ is the range of electron energies passed by the spectrometer. Unless a fixed pass energy is used, $\Delta\epsilon = k_s \epsilon_s$ where k_s is a spectrometer constant determined by spectrometer type and geometry. Thus, the real electron signal is given by:

$$N_s(\epsilon_s, \theta) = c_N \frac{d^2\sigma(\epsilon_s, \theta)}{d\epsilon d\Omega} \Delta\Omega_{out} k_s\epsilon_s. \qquad (4.4)$$

The background signal resulting from electron reflection and secondary emission from the back plate, $N_b(\epsilon_s, \theta)$, depends on the number of electrons hitting the plate, on the probability that they are reflected, R, and on the probability that they are detected. Assuming random scattering, the probability that the scattered electrons are detected is proportional to the solid angle subtended by the exit slit, which we will designate as $\Delta\Omega_b$. The number of electrons having sufficient energy so that they hit the back plate is

$$c_N \int_{\epsilon_{min}}^{\epsilon_{max}} \frac{d^2\sigma}{d\epsilon d\Omega} d\epsilon\, \Delta\Omega_{in}. \qquad (4.5)$$

The solid angle $\Delta\Omega_{in}$ accounts for the fact that the electrons must enter the spectrometer. Thus, the background signal is given by:

$$N_b(\epsilon_s, \theta) = \int\limits_{\epsilon_{min}}^{\epsilon_{max}} \frac{d^2\sigma}{d\epsilon d\Omega} d\epsilon \, \Delta\Omega_{in} \sum R \, \Delta\Omega_b. \tag{4.6}$$

Here the summation indicates that the reflection probabilities and solid angles must be calculated for each point of impact and combined.

To simplify notation here, we abbreviate the integrated cross section by $\int_{\epsilon_1}^{\epsilon_2} \sigma(\epsilon, \theta) \, d\epsilon = \sigma'(\{\epsilon_1, \epsilon_2\}, \theta)$ and the double-differential cross section by $d^2\sigma/d\epsilon \, d\Omega = \sigma''(\epsilon, \theta)$. Hence, the relative number of these scattered background signals with respect to the desired signal is given by:

$$\frac{N_b}{N_s}(\epsilon_s, \theta) = \frac{\Delta\Omega_{in}}{\Delta\Omega_{out}} \frac{\overline{R\Delta\Omega_b}}{k_s} \frac{\sigma'(\{\epsilon_{min}, \epsilon_{max}\}, \theta)}{\epsilon_s \, \sigma''(\epsilon_s, \theta)}, \tag{4.7}$$

where we have replaced the summation over the reflection coefficient and the solid angle of the exit slit with average values for these quantities. Lastly, we use

$$\int\limits_{\epsilon_{min}}^{\epsilon_{max}} \frac{d^2\sigma}{d\epsilon d\Omega} d\epsilon = \int\limits_{0}^{\epsilon_{max}} \frac{d^2\sigma}{d\epsilon d\Omega} d\epsilon - \int\limits_{0}^{\epsilon_{min}} \frac{d^2\sigma}{d\epsilon d\Omega} d\epsilon. \tag{4.8}$$

The first term on the right side is equal to the differential cross section which we abbreviate by $\sigma'(\theta)$. Hence, the relative number of scattered background electrons is given by

$$\frac{N_b}{N_s}(\epsilon_s, \theta) = \frac{\Delta\Omega_{in}}{\Delta\Omega_{out}} \frac{\overline{R\Delta\Omega_b}}{k_s} \left(\frac{\sigma'(\theta)}{\epsilon_s \, \sigma''(\epsilon_s, \theta)} - \frac{\sigma'(\{0, \epsilon_{min}\}, \theta)}{\epsilon_s \, \sigma''(\epsilon_s, \theta)} \right). \tag{4.9}$$

The relative magnitude depends on knowing the specific geometry of the spectrometer but an estimate applicable to parallel plate spectrometers can be obtained from the following information: (i) Typically $\Delta\Omega_{out} \approx 0.5 \, \Delta\Omega_{in}$ for singly focusing spectrometers and $\Delta\Omega_{out} \approx \Delta\Omega_{in}$ for doubly focusing spectrometers, (ii) for moderate resolution spectrometers, $k \approx 0.05$ (iii), $\epsilon_{min} = K \epsilon_s$ where K is a constant depending on the spectrometer design. For a direct reading (applied voltage = analyzed electron energy) spectrometer, $K \approx 2$ for a 45° spectrometer and is ≈ 6 for a 30° spectrometer. For spectrometers where the applied voltage is smaller than the analyzed electron energy, K is smaller.

Directing our attention to the bracketed term, recall that the doubly-differential cross sections are roughly constant for $\epsilon < 10 \, \text{eV}$ and that these cross sections dominate the singly-differential cross section. Therefore the second term in brackets is approximately equal to K and the bracketed term becomes small for $\epsilon > K \epsilon_s$. On the other hand, at lower emission energies, i.e., as $\epsilon \to 0$, N_b/N_s increases as $[\epsilon_s]^{-1}$. This means that the scattered background problems increase strongly as the analyzed electron energy decreases toward zero and that scattered background signals are relatively unimportant above 10 eV. This conclusion agrees with the analysis by *Bernardi* and *Meckbach* (1995), but is in disagreement with the results by *Irby* (1995) whose formalism suffers from dimensionally incorrect expressions.

According to the above equation, the most effective ways of reducing these backgrounds are to reduce the probability of reflections, i.e., decrease R; and to restrict the scattered electrons from passing through the exit slit and being detected, i.e., decrease $\Delta\Omega_b$. However, note that the latter cannot be achieved simply by reducing the size of the exit slit since this also decreases k_s. The above formula also demonstrates that the relative importance of scattered background signals increases dramatically when forward electron emission induced by dressed particle impact is being investigated. This is because projectile ionization contributes significantly to the forward emission cross sections, $d\sigma(\theta)/d\Omega$, i.e., projectile ionization leads to a large number of energetic electrons which strike the back plate and contribute to the scattered background signal.

Experimentally, these scattered background signals are difficult to identify because they are proportional to the target gas density, e.g., simply removing the target gas reduces both the "real" and "background" signals proportionally. Often the relative importance of these backgrounds has been estimated from measurements performed by injecting a monoenergetic source of electrons into the spectrometer and measuring the relative strength of the scattered signals at energies other than the injection energy. Typically, the scattered signal intensity must be at least three orders of magnitude smaller than the injected signal for this type of background problem to be negligible.

Reducing the probability of reflections from the back plate is also effective in reducing the scattered background signals. For this purpose, several methods have been used. The solid back plate of the spectrometer can be replaced with a high transmission grid, or an opening can be cut in the back plate, or the spectrometer surfaces can be coated with carbon soot [*DuBois* (1993, 1994)]. The scattered signals can also be reduced by improving the collimation of the electron trajectories which emerge from the exit slit and are detected. This reduces $\Delta\Omega_b$ without influencing k_s. Note that this collimation method is incompatible with attempts to increase the data accumulation rate by replacing the narrow exit slit of the spectrometer with a large area position sensitive detector.

For electrostatic spectrometers with geometry such that the back plate voltage, V_B, is less than the analyzed electron energy, ϵ, scattered background signals can also be reduced by placing a grid between the spectrometer exit slit and the detector, and biasing the grid with a voltage, V_b, greater than V_B but less than ϵ/e, where e is the electron charge. Thus, electrons with energies less than eV_b will be rejected [*Stolterfoht* (1971)]. In principle, eV_b can be made arbitrarily close to ϵ/e but in practice electron optics limit how large V_b can be.

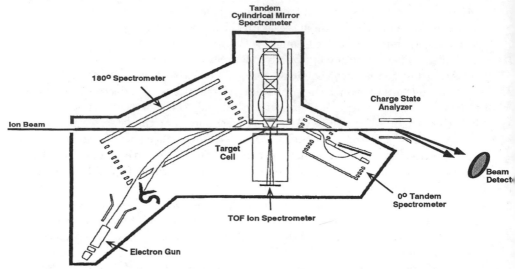

Fig. 4.4. The multispectrometer coincidence apparatus used at Oak Ridge National Laboratory to investigate recoil ion-electron coincidences. From *Gaither* et al. (1993)

Another method of reducing backgrounds is to use two spectrometers operated in tandem, i.e., where the output of the first stage serves as the input for the second stage. Examples of tandem spectrometers are shown in Figs. 4.3 and 4.4. This method, commonly used in zero-degree electron emission studies, was pioneered at the Hahn-Meitner Institute [*Itoh* et al. (1983, 1985)]. Scattered signals produced within the first stage tend to be rejected in the second stage. However, longer trajectories from the source to the detector result, and this increases the possibility that residual magnetic or electrostatic fields may perturb the flight path. It is important to remember that *a priori* predictions about the scattered electron backgrounds cannot be made, but scattered electron backgrounds can seriously influence low-energy electron emission measurements.

4.3.3 High-Energy Electrons

Let us return our attention to the high-energy end of the electron spectrum. These electrons are produced in violent binary collisions. In order to probe these collisions at impact energies higher than those investigated to date would require measuring electron emission above 10 keV. For electrostatic electron energy analysis, this means that the electric field in the spectrometer must be increased, either via geometric factors associated with the spectrometer design or by applying higher voltages to the spectrometer plates. However, the higher voltages can lead to electrical breakdown problems which means that bulky wires and increased isolation of the spectrometer plates from ground are required. Unfortunately, this inhibits the spectrometer movement, and often restricts the angular range for the measurements. Another problem associated with measuring high-energy electron emission is that the ionization cross sections become increasingly smaller, with respect to those for low-energy electron emission. Therefore, extremely good rejection and shielding of background signals is necessary. For these reasons, electron emission above 5 to 10 keV has rarely been investigated.

4.4 Coincidence Techniques

For heavy particle impact, ionization pathways that are nonexistent or negligible for proton impact can make significant contributions to the electron emission. These pathways include ionization from either, or both, centers (in the case of dressed ion impact), multiple electron emission from either center, ionization processes associated with electron transfer between centers, and inner- versus outer-shell ionization processes associated with either center. In some cases, it is possible to isolate and study these various ionization pathways by measuring coincidences between emitted electrons and projectile, or target, ions or coincidences between inner- and outer-shell ionization products. It should be remembered, however, that obtaining such detailed information via coincidence studies has a price, namely, that high resolution studies with small statistical uncertainties cannot be performed. Examples of techniques that have been used, or may be useful, in investigating specific ionization channels follow.

4.4.1 Electron-Projectile Ion Coincidences

Coincidences between emitted electrons and projectile ions can provide information about emission from (a) the projectile center, (b) emission from both centers in a single collision, (c) emission associated with electron transfer between centers, and (d) information about ionization probabilities as a function of impact parameter. Cases (a) and (b) involve coincidences with projectiles that have lost electron(s), case (c) with projectiles that have captured electron(s), and case (d) with projectiles that may, or may not, have changed their charge state in the collision.

A schematic diagram of a typical apparatus used to investigate electron emission associated with electron capture by or loss from the projectile is shown in Fig. 4.5. It differs from a "standard" differential electron emission apparatus in that post-collision projectile-ion charge state analysis and particle detection, plus coincidence electronics, have been added. In studies at the Pacific Northwest Laboratory, [*DuBois* and *Manson* (1986, 1990), *DuBois* (1994)] a surface barrier detector was used to detect the fast projectile ion signal, whereas channeltrons were used at the University of Frankfurt [*Heil* et al.

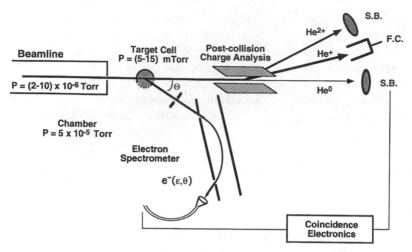

Fig. 4.5. The coincidence apparatus used at the Pacific Northwest Laboratory to investigate ionized electron-projectile ion coincidences. From *DuBois* and *Manson* (1986, 1990)

(1991a,b, 1992), *Trabold* et al. (1992), *Kuzel* et al. (1993a,b)] and ATOMKI [*Kövér* et al. (1989)]. For investigations of the electron emission as a function of the projectile scattering angle [*Schiwietz* et al. (1987), *Jagutzki* et al. (1991a,b)] or charge state [*Quinteros* et al. (1991)], groups at Kansas State University and the University of Frankfurt have used position sensitive parallel plate avalanche detectors to detect the projectile ions.

As an example of typical experimental parameters in these coincidence experiments, the study of electron emission associated with projectile ionization in $He^+ + He$ collisions will be used. This study was performed at the Pacific Northwest Laboratory. The parameters are: spectrometer solid angle and energy resolution, $\Delta\Omega_e \approx 0.3$ mrad and $\Delta\epsilon/\epsilon = 12\%$, target pressure ≈ 15 μbar, He^+ beam intensity ≈ 1 pA, He^{2+} intensity ≈ 150 kcounts/sec, electron count rate at $20°$ ≈ 1 count/sec or less, e^- - He^{2+} coincidence intensity 20–80 % of the electron signal rate. At larger emission angles, the electron signals and, hence, the coincidence rates were smaller.

An important experimental consideration associated with this type of coincidence measurement is to ensure that coincidence signals associated with double collisions have negligible importance. For example, false electron-ionized projectile coincidence signals can be produced if the projectile is ionized before, or after, the target cell and a collision in the target cell produces an electron that is detected. The probability of these double collision events must be small with respect to the probability that a projectile ion and an electron are produced in a single collision in the target cell. These double collision events can be minimized by reducing pressures outside the target cell and shortening the distances between the target cell and the pre- and post-collision charge state analyzers.

The example cited was for coincidences involving the electron-loss channel, but similar signal rates and experimental difficulties occur in studies where the electron emission associated with the electron capture channel is investigated.

4.4.2 Electron-Target Ion Coincidences

Coincidence measurements between emitted electrons and ionized targets can provide information about multiple electron emission from the target. Few measurements of this type have been performed, however, because the coincidence signal rates are extremely small. As an example, let us calculate the coincident rates expected for a test case of interest, namely double ionization of helium resulting from 1 MeV proton impact. We will assume a simple and straightforward experimental setup consisting of a parallel-plate electron spectrometer and a "standard" time-of-flight recoil ion detector. Let us further assume that the electron spectrometer has a modest energy and angular resolution of $\Delta\epsilon/\epsilon = 10\,\%$ and $\Delta\Omega = 0.01$ sr and that the recoil ion detector can handle a count rate of 10^5 s^{-1}.

The maximum recoil ion counting rate limits the projectile beam intensity, N_B, that can be used. It is known from previous studies that even a modest electrostatic extraction field is sufficient to collect all target ions, i.e. $\Delta\Omega_i = 4\pi$; thus the appropriate equation is:

$$N_i = \sigma\, N_B\, N_T\, L\, \Delta\Omega_i/\, 4\pi. \tag{4.10}$$

Here σ is the total ionization cross section, N_i is the recoil ion count rate, N_T is the target density, L is length of the target observed by the recoil ion detector, and $\Delta\Omega_i$ is the recoil ion detection solid angle. For simplicity, we have assumed the ion detection efficiency is 100 %. Let us further assume that the target pressure is 1 μbar and the length of the target where ions are extracted is $L \approx 0.5$ cm. The total ionization cross section is approximately 2×10^{-17} cm^2 [Rudd et al. (1983)]. With these parameters, the beam intensity must be restricted to 3 pA or less in order to avoid swamping the recoil ion detector.

The doubly-differential non-coincidence electron counting rate, N_e, can be determined using:

$$N_e = \frac{d^2\sigma}{d\epsilon d\Omega}\, N_B\, N_T\, L\, \Delta\Omega\, (\Delta\epsilon/\epsilon)\, \epsilon\, \eta, \tag{4.11}$$

where the electron spectrometer solid angle, $\Delta\Omega$, and energy resolution, $\Delta\epsilon/\epsilon$, were given above. Unit electron detection efficiency, η, is assumed, and $d^2\sigma/d\epsilon\, d\Omega$ is the doubly differential cross section for electron emission [Rudd, Toburen and Stolterfoht (1976)]. For a 3 pA beam, a non-coincidence electron counting rate of 0.2 and 0.1 count/sec at 10 and 100 eV respectively is therefore expected at 20°. Of these rates only a small fraction (less than 1 % according to total ionization cross section information (DuBois and Manson, 1987) is expected to be associated with the double ionization events of interest. Thus, a coincidence rate of approximately 1 count every 10 minutes is predicted. Even under optimum conditions, namely heavy ion impact on heavy targets, the counting rates remain formidably small. This is because heavy ion impact increases the target ionization cross section, but this also increases the recoil ion count rate. The net effect is that the beam current must be reduced from the value given above, and therefore the coincidence rate increases very little, if at all.

Clearly, electron-target ion coincidence experiments require unique methods to overcome these extremely low counting rates. Obvious approaches are to increase the signal rates by (a) increasing the solid angle of the electron spectrometer, (b) increasing the band of electron energies sampled at any given time, and (c) filtering out unwanted recoil ions in order to increase the beam intensity.

One method of increasing the solid angle of the electron spectrometer is to use several electron spectrometers, as shown in Fig. 4.4. [*Gaither* et al. (1993)]. In this apparatus, the parallel-plate spectrometers collected electrons in a ± 7.5 angular range in the scattering plane while the cylindrical mirror spectrometer used a position sensitive detector to record electrons emitted between 39° and 144°. A small electric field was used to extract target ions with minimal perturbation on the electrons of interest.

An innovative method for increasing the electron detection solid angle was developed at the University of Nebraska [*Chung* and *Rudd* (1996)] and used to study singly differential cross sections for proton impact on atoms. The basic concept, shown schematically in Fig. 4.6, uses a large hemispherical

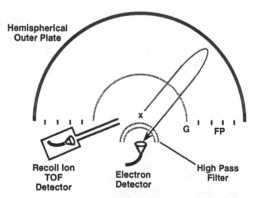

Fig. 4.6. The 2π spectrometer used at the University of Nebraska to investigate recoil ion-electron coincidences. x: interaction point defined by intersection of ion beam and gas jet target; G: hemispherical grid used as inner spectrometer plate; FP: field straightener plates. Two biased, hemispherical grids form the high pass energy filter. From *Chung* and *Rudd* (1996)

electrostatic mirror to focus nearly 2π of the electron emission onto a detector. Emission energy information is achieved by biasing the outer hemisphere and the hemispherical grids to produce a high-pass low-pass energy filter. A small electric field was used to extract target ions.

Another powerful technique for studying electron-target ion coincidences was developed at the Hahn-Meitner-Institut Berlin [*Schiwietz* et al. (1987), *Skogvall* and *Schiwietz* (1990), *Schiwietz* et al. (1994)]. The apparatus used one or two time-of-flight electron spectrometers having a large solid angle of 0.15 radian. This considerably increased the electron counting rates and hence the coincidence signals. Further increases were obtained by using a pulsed target ion extraction field. Normally the extraction field was near zero, but after detection of an electron, a strong extraction field was pulsed on. This pulsed-field method removes strong perturbations that the extraction field might have on the electron trajectory, and helps avoid high recoil ion counting rates since ions associated with electrons which were emitted, but not detected, are neither extracted nor counted. Although very powerful, this combination of time-of-flight electron energy analysis and recoil ion pulsed extraction field method has only been applied to proton impact studies.

Using electrostatic electron energy analysis, researchers at the University of Frankfurt have applied the pulsed target ion extraction technique to heavy ion impact collisions, specifically He^+ impact on neon and argon [*Keller* (1989)]. The interest of these studies was zero-degree cusp electron production. It should be pointed out that pulsed extraction field techniques must be carefully tested to demonstrate that they do not produce false coincidences. For example, if a target ion produced in an earlier collision is still in the interaction region when the field is pulsed on, this "false" ion would be extracted. In certain cases, the "false" ion could be detected and "false" coincidence information would be recorded. Also, that the pulse voltage can introduce false signals in the particle detector amplifiers, unless adequate shielding is used.

As demonstrated by our sample calculation above, the limiting factor in the target ion-electron coincidence counting rate is the very high recoil ion rate. A unique method of reducing the high recoil ion count rate has been used at the University of Frankfurt [*Manzey* (1991)] and is diagrammed in Fig. 4.7. In this case, the recoil ions are extracted from the interaction region by a small electric field but an electrostatic gate keeps them from reaching the recoil ion detector. The gate is opened after an electron is detected but it is opened only for a specific recoil ion charge state. For example, ions extracted from the interaction region have velocities proportional to their charge state. Therefore, they are spatially separated along their flight path and the gate can be opened/closed for the desired/undesired target ions. Thus, the vast majority of the recoil ion signal can be rejected by simply closing the gate for all singly charged target ions. This method has only been exploited in this single experiment but offers great possibilities for future studies.

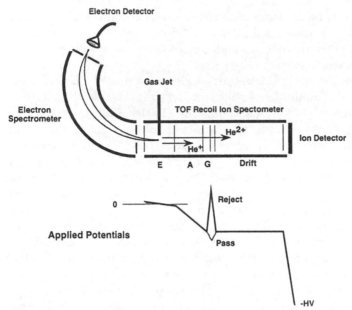

Fig. 4.7. The "electronic gate" spectrometer used at the University of Frankfurt to investigate recoil ion-electron coincidences. The time-of-flight recoil ion spectrometer has several grids that are biased as shown to produce an extraction, E, region, an acceleration, A, region, a gate, G, region, and a field-free drift region. The gate potential is normally set to reject ions but is biased to pass the appropriate charge state ions after an electron has been detected. From *Manzey* (1991)

Finally, at the time this review was being written, recoil ion momentum spectrometers and techniques are being applied to include the investigation of differential low-energy electron emission in coincidence with specific degrees of target ionization [*Moshammer* et al. (1994, 1996)]. As these are relatively new developments and the data are typically presented in terms of momenta components, rather than energies and angles, the reader is referred to publications by *Wu* et al. (1994), *Dörner* et al. (1994), and *Moshammer* et al. (1994, 1996) for additional information.

4.4.3 Outer Shell-Inner Shell Coincidences

The final type of coincidence technique that we will discuss involves coincidences between electrons and relaxation products, either electrons or photons, stemming from inner-shell ionization. The only reported studies of this type have been performed for proton impact on argon, [*Sarkadi* et al. (1983, 1984), *Weiter* and *Schuch* (1982)], but the methods are directly applicable to heavy ion impact. These studies demonstrate that outer shell-inner shell coincidence measurements are subject to extremely low coincidence rates and to poor signal-to-noise ratios. The low rates result from the requirement of two spectrometers, each of which records only a small portion of the emission spectrum because of restricted solid angle and energy resolution whereas the poor signal-to-noise ratios are the result of small inner-shell cross sections relative to outer-shell cross sections.

For example, except for the inner-shell relaxation product that was detected, the basic techniques and signal rates for these experiments were similar. The inner-shell relaxation product, e.g., argon LMM Auger electrons or K x-rays, were measured at, or near, 90° relative to the beam direction. An electron spectrometer with modest energy resolution and relatively large solid angle was used to record the differential electron emission at forward and backward angles. For coincidences involving argon LMM Auger electrons, the coincidence rate was approximately 0.001 count/s. For coincidences involving argon K x-rays, a much higher coincidence rate of 0.01–1 count/sec was recorded. But this higher rate contained a high random coincidence background of 60–80 % of the measured coincidence signal.

From these signal rates, it appears that this technique has limited application in future studies. While it is possible to address the problem of low coincidence rates by improving the experimental design, nothing can be done about the poor signal-to-noise ratios.

5. Ionization by Bare Projectiles

Electron spectra induced by bare heavy ions generally exhibit pronounced structures due to soft and binary collisions (Fig. 2.1). In addition, at observation angles near $0°$, the spectra show the cusp shaped peak due to electron capture to the continuum. The ionization mechanisms for bare heavy projectiles are similar to those for proton impact. The proton impact data have been discussed in detail by *Rudd* et al. (1992). Therefore, in the following, the attention is focused on the characteristic features of bare heavy ions. Contrary to proton impact, highly charged ions give rise to significant two-center effects on the ejected electrons at nearly all emission angles.

The ionization mechanisms associated with heavy ions are discussed in terms of electron emission from charge centers. First, it will be shown that the essential features of the binary-encounter peak, produced by bare projectiles, can be described by theories involving no center or a single projectile center. Then, the discussion is devoted to target centered phenomena. In particular, soft-collision electrons are attributed to a single-center treatment. Although they may be significantly influenced by two-center effects, soft-collision electrons reveal predominant features of the target center. As specific two-center phenomena, electron capture to the continuum and saddle point electrons are treated. Finally, general properties of two-center electron emission at increasing energy are discussed.

5.1 Binary Encounter Collisions

When the interaction of the active electron with the target nucleus is neglected during the collision, the ionization process is determined by a two-body mechanism referred to as binary-encounter electron emission. The characteristic features of the BE mechanism have been introduced in Sect. 2.2. Binary-collisions are usually associated with a single-center formed by the projectile nucleus. The BE mechanism for bare projectiles is depicted in Fig. 5.1a. In the projectile frame, the BE process is analogous to large angle elastic scattering of the electron from the incoming projectile. Classically, the most energetic electrons that can be produced in atomic collisions are these binary encounter electrons.

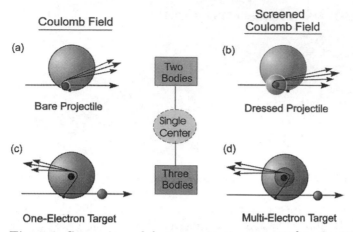

Fig. 5.1. Comparison of electron scattering near 180° in the field of a bare (**a**) and dressed (**b**) projectile nucleus. In (**c**) and (**d**) the corresponding backscattering is shown for the target nucleus. Note that large angle scattering is more probable for the dressed heavy particle than for the bare ones. From *Stolterfoht* (1996)

For an electron initially at rest, the location of the BE maximum is a function of the reduced projectile energy T and electron emission angle θ following from two-body kinematics given by (2.1). Since the electrons are bound to a nucleus, the velocity distribution of the electrons must be accounted for and a broad peak centered at ϵ_{BE} is observed. In Sect. 3.4.1 a more detailed analysis [*Bell* et al. (1983), *Fainstein* et al. (1991)] using a free outgoing electron in conjunction with the Born approximation was discussed, yielding (3.88). This expression is recalled here with a somewhat different notation,

$$\epsilon_{BE} \approx 4T\cos^2\theta - 2E_b. \tag{5.1}$$

It is identical to (2.1) except that ϵ_{BE} is reduced by twice the ionization energy E_b. This reduction originates from the initial velocity distribution of the electrons.

The binary-encounter peak was clearly recognized in the early work by Rudd and collaborators [*Rudd* and *Jorgensen* (1963), *Rudd* et al. (1966a)] studying electron emission by proton impact of a few hundred keV. Similar work has been performed at higher projectile energies by *Toburen* (1971) and *Stolterfoht* (1971). These early studies dealing with H^+ + He collisions cover the projectile energy ranges from 5 keV to 5 MeV [*Rudd* et al. (1976)]. For the lowest projectile energies from 5 to 50 keV the binary-encounter electrons do not produce noticeable structures in the electron spectra. As the incident energy increases, the binary encounter maximum develops into a sharp maximum (Fig. 2.5). For the incident energy of 5 MeV the top of the BE maximum is more than an order of magnitude higher than the "background". It should be noted, however, that early experiments were not subject to explicit studies of the binary-encounter profile.

5.1.1 Shape of the Binary Encounter Peak

Detailed studies of the binary encounter peak were initiated by Bell and collaborators in the early 1980s [*Bell* et al. (1983), *Böckl and Bell* (1983)]. They have shown that the BE peak shape is determined by the Compton profile which, in turn, is governed by the initial velocity distribution of the electron that is ejected. The analysis of the BE peak profile has been performed by means of the free-electron peaking approximation (FEPA) describing the zero-center case (Sects. 3.3.4, 3.4.1). Thus, for the double-differential cross section, *Bell* et al. (1983) provided the simple formula (3.91) which shall be rewritten here

$$\frac{d\sigma}{d\epsilon \, d\Omega} = \frac{4Z_p^2}{v^2 \, k^3} J(p_z), \tag{5.2}$$

where $J(p_z)$ is the Compton profile determined by the z component of the initial momentum

$$p_z = k \, \cos\theta - \frac{\Delta\epsilon}{v} = k \, \cos\theta - \frac{k^2}{2v} - \frac{E_b}{v}. \tag{5.3}$$

It is recalled that θ and k are the angle and momentum of the ejected electron, E_b is its ionization energy, v is the projectile velocity, and $\Delta\epsilon$ is the energy transfer. The Compton profile $J(p_z)$ for a hydrogenic orbital may be obtained in closed form, e.g., (3.13). Results for multielectron systems are tabulated in the literature by *Biggs* et al. (1975)

In Fig. 5.2 the results of this FEPA model are compared with experiments for 1-MeV H$^+$ + He collisions [*Böckl* and *Bell* (1983)]. The data for a variety of emission angles are plotted as a function of the momentum component p_z. In this type of plot the differential cross sections for the different angles are seen to coincide within the experimental uncertainties. It is noted that the model calculations underestimate the low energy wing of the BE peak. Otherwise, the theory is in good agreement with experiments showing that the FEPA model reproduces the essential features of the binary-encounter peak.

Recently, new interest in the shape of the binary-encounter peak has been generated since detailed measurements of continuous electron spectra have been performed at the observation angle of 0° [*Lee* et al. (1989, 1990)]. To interpret the experimental data, *Lee* et al. (1990) used a theoretical approach which is similar to that of the binary-encounter theory. The model involves an impulse approximation which adopts a free incident electron, liberated from the target before the collision. Hence, Lee et al. (1990) refer to their theoretical model as the impulse approximation (IA). The analysis is described in the theoretical Sect. 3.1.2 resulting in a relative simple expression given by (3.11).

When the incident ion carries electrons, more effort is required to determine the elastic scattering cross section $d\sigma'/d\Omega'$. This case involving a single

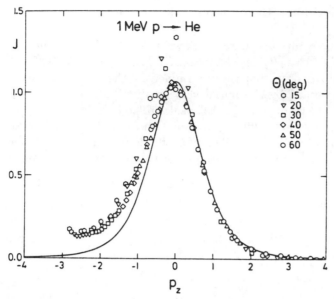

Fig. 5.2. Compton profile of He for electrons ejected at different angles θ by 1-MeV H^+ impact. The experimental data are from the compilation by *Rudd* et al. (1976). The Compton profile is based on data tabulated by *Biggs* et al. (1975). The figure is taken from *Böckl* and *Bell* (1983)

projectile center will be discussed in a later section devoted to dressed projectiles. For bare projectiles, treated here, $d\sigma'/d\Omega'$ is equal to the Rutherford cross section $d\sigma'_R/d\Omega'$ that is given in closed form by (3.1). Hence, inserting that cross section into (3.11) one obtains a convenient expression for the double-differential electron emission cross section

$$
\frac{d\sigma}{d\Omega \, d\epsilon} = \sqrt{\frac{\epsilon}{\epsilon'}} \frac{Z_p^2}{16 \, \epsilon'^2 \, \sin^4\left(\theta'/2\right)} \frac{J(p_z)}{\sqrt{2T} + p_z},
\tag{5.4}
$$

where $J(p_z)$ is again the Compton profile. Note that this expression describes the zero-center case.

The electron energy ϵ' and emission angle θ' in the projectile frame of reference may be obtained from the corresponding quantities ϵ and θ in the laboratory reference frame using the kinematic transformation rules given in Appendix B. Here, the interest is focused on the laboratory observation angle of $\theta = 0°$ which corresponds to $\theta' = 180°$ in the projectile frame of reference. In this case the kinematic transformation is achieved by

$$
\epsilon' = \left(\sqrt{\epsilon} - \sqrt{T}\right)^2.
\tag{5.5}
$$

It is recalled that T is the reduced projectile energy.

A comparison of experimental and theoretical data is shown in Fig. 5.3. The data are measured by *Lee* et al. (1990) using different projectiles inci-

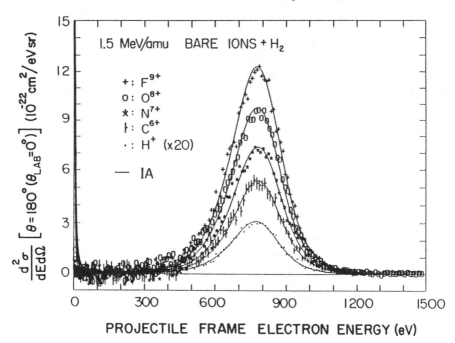

Fig. 5.3. Binary-encounter peaks produced by five different bare projectiles. Typical statistical errors are shown for C^{6+}. Beam induced background is subtracted. The experimental data are compared with calculations using the IA model. The data for proton impact are multiplied by a factor of 20. From *Lee* et al. (1990)

dent on H_2. The experimental data are normalized to the theoretical results by one common factor. It is seen that their IA model is in excellent agreement with experiment. Both the experimental peak profiles and the relative intensities are accurately reproduced by the theory. A striking result of the comparison between theory and experiment is the precise confirmation of the Z_p^2 dependence of the BE cross section predicted by (5.4). *Lee* et al. (1990) have shown explicitly that the integrated BE cross sections divided by Z_p^2 are constant within the experimental uncertainties. A further test of the theory has been performed with regard to the projectile energy dependence of the binary-encounter cross section. *Lee* et al. (1990) have verified that the cross section at the top of the binary-encounter maximum follows a $Z_p^{2.6}$ law for 19- to 38-MeV F^{9+} impact on H_2 in accordance with the theoretical prediction.

The remarkable agreement between the IA model and experiment calls for a comparison with other theoretical methods. In Fig. 5.4 the experimental data by *Lee* et al. (1990) are shown together with different theoretical predictions for the impact of 1.5 MeV/u F^{9+} on He. The theoretical results [*Schultz* and *Reinhold* (1994)] correspond to the use of the CTMC and CDW-EIS approximations described in Sects. 3.2, 3.5.1, respectively. CDW-EIS calculations have been carried out using a model potential [*Garvey* et al. (1975)]

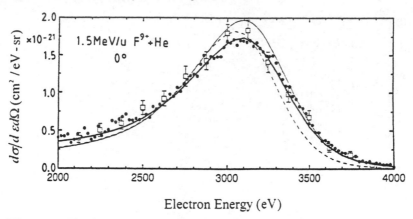

Fig. 5.4. The binary-encounter peak region for electrons ejected at $0°$ in collisions of $1.5\,\mathrm{MeV/u}$ F^{9+} with helium. *Solid circles:* experimental data of *Lee* et al. (1990), *open squares:* CTMC results, *solid line:* CDW-EIS calculations using effective Coulomb target fields, *heavy solid line:* model potential CDW-EIS calculations; *dotted line:* B1 results, and *dashed line:* DSPB predictions The DSPB results are from *Brauner* and *Macek* (1992) and the other theoretical data are from *Schultz* and *Reinhold* (1994)

to compute the initial bound and final continuum target wave functions or effective target Coulomb fields of charges $Z_t^i = 1.68$ and $Z_t^f = 1.34$. The CTMC results (Sect. 3.2) are performed using a microcanonical distribution in which the interaction of the active electron with the residual target core is described by the model potential used for the CDW-EIS model. DSPB calculations obtained for initial and final Coulomb target fields with effective charges $Z_t^i = Z_t^f = 1.34$ [*Brauner* and *Macek* (1992)] are also shown.

For a further examination of the analytic models it is instructive to also compare with the Born approximation. In Fig. 5.5, the results IA and the FEPA models are plotted together with data obtained from the first Born approximation (B1) described in Sect. 3.4.2. The calculations by means of the IA and FEPA contain (scaled) hydrogenic bound state wave functions for both H and He. It is mentioned that *Zouros* et al. (1994) used more realistic Compton profiles for He yielding IA results in good agreement with the present ones. The B1 results were evaluated using a negligibly small projectile charge in the CDW code by *Gulyás* et al. (1995a). They involve Hartree-Fock-Slater wave functions for He.

First, we consider the H target in Fig. 5.5b where excellent agreement is achieved between the B1 results and the other model calculations. [The theoretical data also coincide with the experimental data reported by *Zouros* et al. (1994).] The finding that the Born approximation correctly describes the production of the binary-encounter peak appears to be surprising. Binary encounter electrons are created in violent collisions for which perturbation theory is not expected to be valid. To explain this seeming controversy, it is re-

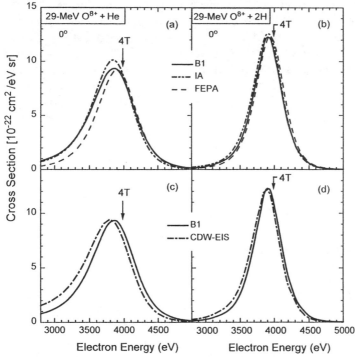

Fig. 5.5a–d. Comparison of different models used to calculate cross sections for the production of the binary-encounter peak by 29-MeV O^{8+} impact on 2 H atoms and He. The FEPA and IA model results are obtained with scaled hydrogenic wave functions. The B1 and CDW-EIS include Hartree-Fock-Slater wave functions for He

called that the two-body Coulomb problem yields the same result when solved either in first- or higher-order perturbation theory (Madison and Merzbacher, 1975). This should be considered a mere accident for bare projectiles treated here, as it does not occur for ions carrying electrons (Sect. 3.7.2).

Regarding the He target, Fig. 5.5a shows that the low-energy part of the BE maximum from the IA model is enhanced in comparison with that of the FEPA. As the FEPA underestimates the experiment in this range (Fig. 5.2) the enhancement by the IA model improves the agreement between theory and experiment. Nevertheless, the different results by the IA and FEPA are noteworthy as similar assumptions are incorporated into the two models. It is recalled that both models are based on a free incident electron picture. The partial neglect of the initial momentum (peaking approximation) in the FEPA model is similar to the neglect of the initial transverse momentum in the IA model, (3.9). Hence, the results from the IA and FEPA are expected to be similar.

The enhancement of the IA curve is primarily caused by the procedure of subtracting the ionization energy E_b from the electron energy (3.9). The

Rutherford scattering is enhanced for low-energy electrons due to the ϵ'^{-2} term in (5.4). Note that the FEPA has a weaker dependence, i.e. by $k^{-3} = \epsilon^{-3/2}$. The difference between IA and FEPA is particularly strong for He, as the corresponding Compton profile is relatively large. It should be added that the ionization energy subtraction from the incident energy gives rise to an energy shift of the BE peak which is discussed in the next section.

Before that discussion we point out a useful property of the binary encounter peak that can be directly utilized in the experiments. The excellent agreement between theory and experiment provides confidence that the binary encounter peak can be used *in situ* to normalize experimental cross sections with relatively high precision. Absolute values for electron production cross sections are usually determined with typical errors of about 30%. It is likely that the uncertainty of the binary-encounter normalization method is smaller than this experimental uncertainty.

5.1.2 Shift of the Binary Encounter Peak

The energy shift of the binary-encounter peak has been the subject of considerable current interest. From Fig. 5.5 it is evident that the BE maximum is located at energies lower than $4T$ predicted by the simple two-body formula (2.1). This energy shifts with respect to the predicted $4T$ value occur for all model calculations. It may be attributed to the initial momentum distribution which is essentially taken into account by the subtraction of $2E_b$ in (5.1). However, closer inspection of the theoretical data shows that even this simple model yields energy shifts which are larger than the $2E_b$ value. This can clearly be demonstrated when the peak position is accurately determined by differentiating the binary-encounter curves.

Different effects are responsible for additional peak shifts associated with the different theoretical models. For instance, the B1 maximum is shifted to lower energies with respect to the FEPA curve (Fig. 5.5a). This shift can be associated with a target-center effect involving the scattering at the target nucleus. It should be realized that the target scattering leads to a breakdown of the impulse approximation. When the electrons, initially ejected at 0°, are scattered by the target atom, a fraction of electrons with the highest possible energy are lost. [Recall from (5.1) that electrons initially ejected at 0° have the maximum energy.] On the other hand, electrons initially ejected at angles larger than 0° and later rescattered to 0° by the target will increase the contribution of lower-energy electrons. Hence, the backscattering in the target center will reduce the overall energy of the binary encounter peak observed at 0°.

Similar to the B1 data, the IA curve exhibits an energy shift which is clearly visible for the He target (Fig. 5.5a). This shift is primarily caused by subtracting the ionization energy E_b from the energy of the ejected electron (3.9). It should be recalled that the IA model yields excellent agreement with experiment (Fig. 5.3). This agreement is as good as or superior to that of the

Born approximation [*Lee* et al. (1990)]. We note that a one-center theory such as the B1 approximation should be more accurate than the simplification of an outgoing plane wave for the electron. Evidently, the subtraction of E_b in the IA model mimics other effects causing energy shifts.

A further energy shift is produced by two-center effects influencing the shape and the position of the binary encounter peak [*Fainstein* et al. (1992)]. Figures 5.5c,d show the comparison of the results from B1 and CDW-EIS calculations. Again for the H target, the model results agree quite well, whereas for He the CDW-EIS curve is significantly shifted with respect to the B1 curve. *Fainstein* et al. (1992) have demonstrated in various examples that the CDW-EIS model is in good agreement with experiment. As the CDW-EIS model accounts for two-center effects, it is likely that these effects are responsible for the further energy shifts of the binary-encounter peak.

The reason for the energy shift in the CDW-EIS calculation is not fully understood at present. According to a model by Bohr and Lindhard [*Pedersen* et al. (1991)] it is unlikely that the electron is liberated before the collision (as assumed by the impulse approximation). The electron is ionized at a finite internuclear distance when the projectile potential becomes noticeable. The attractive field of the projectile cannot transfer as much energy to a bound (passive) electron as it can to a free (active) electron, see also Fig. 2.6. Thus, with respect to a free-electron model, kinetic energy is missing when the electron separates from the projectile. Similar arguments have been incorporated into the model of *Fainstein* et al. (1992). In addition, we recall a characteristic two-center effect, i.e., the deflection of the electron in the projectile field (Fig. 2.2). Hence, lower-energy electrons initially ejected at non-zero angles may be deflected in the projectile field and, thus, reach the detector at 0°. A similar effect was noted above for scattering in the target-center. It should be pointed out that shifts to lower energies, originating from electron deflection, are particular important for observation angles of 0°.

Recently, several zero-degree measurements were carried out revealing detailed information about the energy shift of the binary-encounter peak [*Lee* et al. (1990), *Pedersen* et al. (1991), *Hvelplund* et al. (1991), *Hidmi* et al. (1993), *González* et al. (1992), *Sataka* et al. (1993, 1994)]. Earlier measurements (Fig. 2.5) that were performed at larger emission angles indicated significant energy shifts of the BE peak, however, it was difficult to compare the experimental data with theory. We recall from (5.1) that the BE energy depends on the reduced beam energy T and on $\cos^2\theta$. Experimental uncertainties, primarily in T, precluded sufficiently accurate determinations of BE in order to observe the small energy shifts. However, for zero-degree emission the position of the ECC cusp can be used to precisely determine T and small uncertainties in the emission angle produce negligible changes in $\cos^2\theta$. Thus, using zero-degree electron spectroscopy, it is possible to verify that the binary encounter peak is not located where it is expected but is shifted to lower energies.

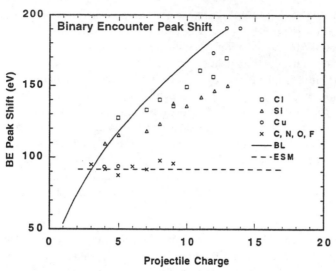

Fig. 5.6. Experimental and theoretically predicted shifts in the location of the zero-degree binary encounter peak as a function of projectile charge state. The experimental data are from Hidmi et al. (1993) with the x being averages of the C, N, O and F data. The *dashed* and *solid curves* are theoretical predictions based on an elastic scattering and the Bohr-Lindhard model, as described in the text

Hidmi et al. (1993) investigated these shifts in detail and found a constant value for all charge states of low Z_p, i.e., $Z_p < 10$, ions as shown in Fig. 5.6. For low Z_p projectiles, the measured shifts are compatible with the elastic scattering model which predicts a constant energy shift [*Bhalla* and *Shingal* (1991)]. In this case it is expected that one-center effects dominate. Heavier ions demonstrated a different behavior, i.e., the energy shift increased with Z_p and/or with q, also shown in Fig. 5.6 [*Hidmi* et al. (1993)]. In this case, two-center effects are expected to gain importance.

Qualitative, but not quantitative, features of these energy shifts can be predicted by theory, as indicated in Fig. 5.6. For higher Z_p projectiles, the Bohr-Lindhard model [*Pedersen* et al. (1991)] predicts energy shifts that increase as $Z_p^{1/2}$ for bare projectiles; experiment indicates a slightly faster increase. The Bohr-Lindhard model is based on the simple picture where the bound target electron is ionized when the target-projectile internuclear separation is such that the Coulomb potential of the projectile matches the binding energy of the target electron. A $Z_p^{1/2}$ dependence is also predicted by the model calculations of *Fainstein* et al. (1992). A consequence of the BE peaks shifting to lower energies with increasing projectile Z_p (or q) is that more higher-energy electrons are emitted in collisions involving low charge state projectile ions rather than in collisions involving higher charge state ions.

In summary, the experimental and theoretical studies revealed different effects that are responsible for the energy shift of the binary-encounter peak. In the foregoing discussion an attempt was made to classify these effects in terms of collision centers. It should be kept in mind that the binary-encounter process is essentially a one-center phenomenon of the projectile. Nevertheless, for bare projectiles, the gross features of the binary-encounter peak are well described by the FEPA model involving no center. For this case, *Bell* et al. (1983) and *Fainstein* et al. (1991) have derived an energy shift that amounts to $-2E_b$ (5.1). The IA model implies a larger energy shift that may mimic one- and two-center effects.

These latter effects are associated with a breakdown of the free-electron (or impulse) approximation. The one-center effect of target backscattering enhances the amount of lower-energy electrons giving rise to a further energy shift. This effect is adequately described by the Born approximation if realistic wave functions are used for the continuum state. It increases with increasing target charge as can be seen when the results for H_2 and He target are compared. Finally, two-center effects, described by the CDW-EIS and CTMC, are found to depend on the projectile charge. They are produced by energy losses of the electron due to ionization at finite internuclear distances and the deflection of the electron in the projectile field.

5.2 Target-Center Phenomena

The foregoing discussion of the binary-encounter peak was based on the assumption that the outgoing electron is free. This impulsive approximation breaks down when the outgoing electron is significantly influenced by the field of the target nucleus, e.g., when the target forms a center (Fig. 2.2). The influence of the target nucleus increases with decreasing electron energy. Therefore, the formation of the target center is particularly important for soft-collision electrons (Fig. 2.3c). Several properties of the soft-collision electrons are contrary to those of the binary-encounter electrons. Binary collisions involve the projectile as a center whereas the soft collisions imply a target center. Moreover, binary collisions are associated with a large momentum transfer whereas in soft collisions the momentum transfer by the projectile reaches its lower limit. Accordingly, binary-encounter collisions may transfer high angular momenta, i.e., values from $l = 50$ to 100 are not unlikely for the data shown, e.g., in Fig. 5.5. In contrast, soft collisions produce dipole transitions that involve the transfer of a unit of angular momentum only.

5.2.1 Early Observations

At the low-energy limit, the interaction between the active electron and the projectile can be considered as a small perturbation which justifies the application of the Born approximation. The wave nature of the outgoing low-energy electron dominates and, therefore, quantum mechanical methods are preferable for the description of the ionization process. In particular, *Bethe* (1930) showed that low-energy electrons are produced in dipole-type transitions similar to those induced by photons. This similarity has been used in various models to determine cross sections for low-energy electron emission [*Kim* (1983), *Rudd* et al. (1992)].

When the energy transfer to the electron increases, scattering in the target center becomes less probable. However, it is still significant for scattering into backward angles (Fig. 2.3d) where electron production by other mechanisms is weak. Naturally, the backscattering becomes more effective as the Coulomb field of the target increases. Hence, in a multielectron target (e.g., Ar), backscattering is expected to be considerable. The different cases of backscattering in single and multielectron targets are depicted in Fig. 5.1c,d.

The analysis of the electrons ejected at backward angles dates back to the pioneering measurements of *Rudd* et al. (1966a). Figure 5.7 shows angular dependencies for electrons ejected by 300-keV H$^+$ impact on He in comparison with calculations using the Born approximation. Angular distributions are given for different electron energies within the range of 10 eV to 750 eV. For energies above $\sim$ 100 eV the data exhibit a maximum at intermediate angles between 50° and 80° that can be identified as the binary-encounter peak. In the region of this peak, good agreement is achieved between experiment and theory. The experimental results show a clear enhancement at forward angles which can be associated with two-center effects where the outgoing electrons are attracted by the receding projectile to forward angles [*Oldham* (1967)]. At backward angles, the experimental data strongly exceed the theoretical results. This finding is attributed to the deflection of the active electron by the target nucleus, see Fig. 2.3d.

In the following, phenomena associated with electron scattering in the target-center will be treated. Emission of soft-collision electrons and backscattering of high-energy electrons in the target center will be separately discussed. Particular attention will be devoted to the analysis of different multipoles.

5.2.2 Energy Distribution of Soft-collision Electrons

Since the beginning of the studies devoted to electron emission in ion-atom collisions, low-energy electrons have received much attention. Here, low energies refer to a few eV with an approximate upper limit equal to the ionization energy of the active electron. The simple reason for this particular interest is due to the fact that the production of low-energy electrons is dominant. These electrons originate from glancing or soft collisions which involve the most probable interactions. Hence, soft-collision electrons govern the total ionization cross sections. Furthermore, in many applications, the production of low-

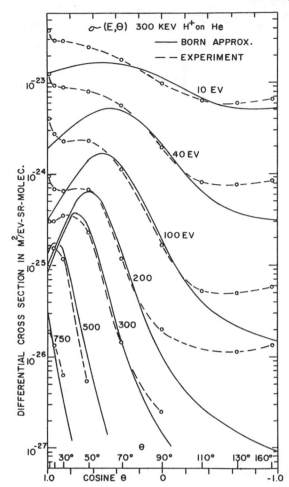

Fig. 5.7. Angular distributions of electrons ejected in 300-keV H^+ + He collisions. The *dotted lines* represent the experiment and the *solid lines* are due to calculations using the Born approximation with scaled hydrogenic wave functions. The electron energy is indicated at the *curves*. From *Rudd* et al. (1966a)

energy electrons play a decisive role. Therefore, the accurate determination of cross sections for soft-collision electrons is of considerable importance.

Unfortunately, the measurements of low-energy electrons below about 10 eV are hampered by several instrumental effects, such as electric and magnetic fields affecting the electron trajectories. These effects have been discussed in detail in the experimental Sect. 4.3.1. When measured, the low-energy electron yields are associated with large experimental uncertainties. Typically, 5 eV and 1 eV electron yields are subject to errors of 30% and 50%, respectively. Because of these uncertainties, consistency checks by means of semi-empirical models have been developed for proton impact [*Kim* (1983)].

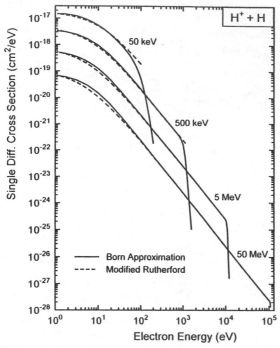

Fig. 5.8. Single-differential cross sections for electron emission in $H^+ + H$ collisions at impact energies of 50 keV, 500 keV, 5 MeV and 50 MeV. The data are calculated using the modified Rutherford formula (7) and the Born approximation formula by *Landau* and *Lifschitz* (1958)

With some caution the same checks may be applied to electron production by heavier bare projectiles. We shall not enter further into the details of the semi-empirical methods as they have recently been described by *Rudd* et al. (1992).

We point out a few general properties of the soft-collision electrons produced by bare projectiles that can be extracted from the modified Rutherford formula (3.5) given in Sect. 3.1.1. As noted earlier, the Rutherford cross section does not take into account the dipole transitions governed by small momentum transfers. However, with the empirical modification of (3.5), we expect a reasonable description of the soft-collision electrons. Examples are given in Fig. 5.8 where results from (3.5) are shown in comparison with corresponding expression (3.99) from the Born approximation for the $H^+ + H$ system [*Landau* and *Lifschitz* (1958)]. The calculations are performed using projectile energies that vary by three orders of magnitude. The agreement is remarkably good in view of the simplicity of the modified Rutherford formula. We note also for the data in Fig. 5.8 that the total cross sections obtained from (3.7) and the Born approximation deviate by less than 20%.

For the following analysis which uses (3.5) it is emphasized that qualitative aspects are more important than quantitative results. First, the number of electrons ejected in a given energy interval can be estimated by integrating (3.5) with respect to ϵ. It is found that about $\sim 65\%$ of the electrons have a kinetic energy lower than the ionization energy, i.e., $\epsilon < E_b$. This clearly shows that low-energy electrons provide the major contribution to the total number of ejected electrons.

Second, we verify the properties of the cross sections at asymptotically small electron energies. The important result of (3.5) is that the single differential cross section increases monotonically and it reaches a finite value as the electron energy ϵ tends to zero (Fig. 5.8). This is an essential property of the Coulomb potential that implies a smooth transition between low-lying continuum states and the adjacent Rydberg series [Burgdörfer (1984)]. Hence, the ionization cross section for small electron energies is determined by the total energy transfer including the binding energy. Therefore, the binding energy is the essential scaling parameter for the low-energy electrons. In particular, it is important to keep in mind that for electron energies $\epsilon \ll E_b$ the differential cross section does not vary much.

Equation (3.5) can be used for a convenient verification of experimental data concerning electron emission at a few eV. First of all it is recalled that the single-differential cross section increases monotonically with decreasing electron energy. Hence, from a rapid drop of measured cross sections at small electron energies it may be concluded that the low-energy electron are influenced by spurious instrumental effects. Similarly, instrumental effects are expected when the cross sections increase dramatically with decreasing electron energy. Such spurious effects may be produced if the electrons are accelerated prior to entering the electron spectrometer. As a reasonable estimate for He, we note from (3.5) that $d\sigma/d\epsilon$ increase by a factor ~ 2 when the electron energy decreases from $\epsilon = 10\,\text{eV}$ to $\epsilon = 1\,\text{eV}$. This factor of ~ 2 increase is in fair agreement with the experimental data given in the cross section compilation by Rudd et al. (1976). From (3.5) it follows that at $\epsilon = 1\,\text{eV}$, the asymptotic limit of $d\sigma/d\epsilon$ is nearly reached for He.

We emphasize that no attempt is made here to predict the cross sections for soft-collision electrons with great accuracy. Due to the uncertainty of the parameter c in (3.5) the extrapolation to zero electron energy remains approximate. Nevertheless, the large experimental uncertainties generally associated with slow electrons justifies using an approximate prediction at the low-energy limit. It is also recalled that the present analysis holds for bare projectiles and single-differential cross sections. With some caution the present method may also be applied to double-differential cross sections. However, it should be kept in mind that the latter cross sections are subject to additional variations at threshold as discussed in the following section.

5.2.3 Angular Distribution of Soft Collision Electrons

Dipole Approximation at High Projectile Energies. It was pointed out earlier that the low-energy electrons are produced by dipole-like transitions, similar to those induced by photons. It is well known from photoionization that dipole transitions produce electrons with a broad angular distribution. In fact, the angular distribution has a $A + B \cos^2 \theta$ like dependence that is symmetric around 90° [*Berkowitz* (1979)]. Similar angular distributions are expected for charged particle impact, when the ionization process is limited to dipole transitions [*Inokuti* (1971)]. However, monopole or quadrupole transitions cannot be ruled out for soft collisions, in particular, at lower incident velocities.

The different contributions of the multipole transitions are incorporated in the expansion (3.108) based on Legendre polynomials P_j [*Meckbach* et al. (1981), *Burgdörfer* et al. (1983)]:

$$\frac{d\sigma}{d\epsilon \, d\Omega} = \sum_{j}^{\infty} b_j \, P_j(\cos\theta), \qquad (5.6)$$

where b_j are expansion coefficients that are independent on the emission angle θ. Within the framework of the Born approximation, the b_j's are obtained as functions of radial matrix elements weighted by Clebsch Gordon coefficients (3.108). The important property of the expansion (5.6) is that incoherent sums of multipole terms are represented by even values of the index j whereas the interferences between multipole terms of different parities are represented by odd values of j.

For instance, the first three terms of (5.6) with $j = 0, 1$, and 2 refer to monopole transitions, monopole-dipole interferences, and dipole transitions, respectively. The dipole approximation performed by taking only these three terms has been discussed in Sect. 3.3.3. The analysis yields (3.71) which allows explicitly treating magnetic quantum numbers. After integration over the impact parameter it follows that

$$\frac{d\sigma}{d\Omega \, d\epsilon} = Q_{0,0} + Q_{1,0} \cos^2 \theta + Q_{1,\pm 1} \sin^2 \theta + A_{1,0}^{0,0} \cos \theta. \qquad (5.7)$$

The quantity $Q_{0,0}$ is the cross section for monopole transitions which is obtained by the impact-parameter integration of the squared matrix element $|I_{0,0}|^2$ from (3.71). Similarly, $Q_{1,0}$, and $Q_{1,\pm 1}$ are obtained as cross sections for dipole transitions involving the magnetic quantum numbers 0 and ± 1, respectively. The asymmetry parameter $A_{1,0}^{0,0}$ describes the interference between the monopole and dipole term.

First, assuming a negligible monopole-dipole interference, it follows from (5.7) that a statistical production ($Q_{1,0} = Q_{1,\pm 1}$) of the magnetic quantum number 0 and ± 1 leads to an isotropic distribution of the electrons. Due to the $\cos^2 \theta$ term, one may conclude that the production of the magnetic quantum number $M = 0$ dominates if the angular distribution exhibits a minimum at 90°. On the contrary, if a maximum at 90° is encountered, the quantum numbers $M = \pm 1$ dominate due to the $\sin^2 \theta$ term. Moreover, the contribution from monopole transitions, added incoherently, provides a constant value which cannot be distinguished from the dipole transitions.

However, the monopole and dipole terms, added coherently, give rise to an asymmetry of the angular distributions around 90°. Thus, monopole transitions (or transitions due to other even multipoles) may be verified experimentally. We note that (5.7) can also be derived from the Born approximation treated in Sect. 3.4.2. Hence, interferences between monopole and dipole transitions, leading to asymmetric angular distributions of the soft-collision electron are possible within the framework of the Born approximation.

Apart from this monopole-dipole interference within the Born approximation, the angular distributions may be influenced by two-center effects. As discussed in Sect. 2.4, two-center effects result in a redistribution of the electron intensity from backward to forward angles so that a possible asymmetry in the angular distribution of the electrons is enhanced. Within the framework of the CDW-EIS model applied for a hydrogenic target, *Cravero* and *Garibotti* (1994) derived analytic expressions for the first 3 terms in (5.7) accounting for an enhanced forward emission of soft-collision electrons. To verify these effects for a He target, we compare theoretical results from the B1 approximation with those from the CDW-EIS with Hartree-Fock-Slater wave functions for the description of the initial and final states. The theoretical results were again obtained using the program by *Gulyás* et al. (1995a).

Figure 5.9 shows measured angular distributions of 2 eV electrons ejected from He in collisions with 5-MeV H^+ and 5-MeV/u Ne^{10+} projectiles of equal-velocity reported by *Rudd* et al. (1976) and *Stolterfoht* et al. (1995), respectively. The experimental results are compared with theoretical data obtained by means of the CDW-EIS and the B1 approximation. First, we focus on the case of H^+ impact in Fig. 5.9a. We note the good agreement between the theoretical results from the CDW-EIS and B1 approximation which indicates that two-center effects are unimportant. Furthermore, the experimental results agree fairly well with both sets of theoretical data. In particular, the angular distribution of the electrons is nearly symmetric around 90°. According to the previous discussion, the cross section maximum at 90° indicates a preferable production of the magnetic quantum numbers ± 1. This finding suggests that the production of 2-eV electrons in 5-MeV H^+ on He collisions is primarily due to dipole transitions. Thus, the 5-MeV H^+ data show that fast protons produce symmetric angular distributions similar to those known from photoionization.

The picture changes considerably when projectiles with higher incident charge state are used. In Fig 5.9b experimental and theoretical data of angular distributions are compared in the case of 5-MeV/u Ne^{10+} impact. Apart from an overall scaling factor $Z_p^2 = 100$, the B1 results are identical to those for 5-MeV H^+ impact in Fig. 5.9a. However, for 5-MeV/u Ne^{10+} impact,

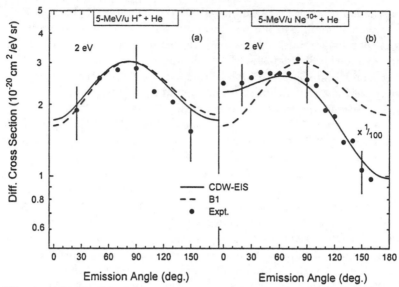

Fig. 5.9a,b. Double differential cross sections for the emission of 2 eV electrons by 5 MeV/u H^+ and Ne^{10+} impact on He. The error bars refer to absolute uncertainties. In (**a**) the experimental H^+ data are from *Rudd* et al. (1976) and in (**b**) the Ne^{10+} data are from *Platten* et al. (1987) and *Stolterfoht* et al. (1995). The theoretical CDW-EIS data were evaluated using the code by Gulyás et al. (1995a). The results due to the Born (B1) approximation were also obtained from that code using a negligibly small projectile charge. The calculations are based on Hartree-Fock Slater wave functions for the initial and final continuum state of the electron. Note that all Ne^{10+} results have been divided by 100

the experiment and the CDW-EIS calculations differ notably from the B1 results. This indicates that two-center effects on soft-collision electrons can be observed for fast and highly charged projectiles. It will be shown in Sect. 5.3.2 that the two-center effects increase strongly with increasing electron energy so that the one-center aspect is lost. For low-energy electrons, however, it is likely that the target center retains a certain importance.

The importance of two-center effects may be estimated by means of the Bohr parameter Z_p/v which is a measure for the perturbation strength of the projectile. (The Bohr parameter Z_p/v will also be used to verify saturation effects in Sect. 5.4.) Two-center effects are expected to be small for $Z_p/v \ll 1$ and become noticeable for $Z_p/v \approx 1$ [*Schneider* et al. (1989), *Fainstein* et al. (1991), *Cravero* and *Garibotti* (1994)]. For the examples in Fig. 5.9 it is noted that 5-MeV/u H^+ and Ne^{10+} are associated with the perturbation strengths $Z_p/v = 0.07$ and 0.7, respectively. These numbers confirm the conclusion that two-center effects are small for 5-MeV/u H^+ but significant for 5-MeV/u Ne^{10+} impact. However, the Z_p/v parameter should be used with care, as quantitative conclusions for soft-collision electrons considered here are not expected to be valid for two-center effects at higher electron energies.

As shown in the next section, two-center effects increase considerably with increasing energy of the ejected electrons.

The data in Fig. 5.9 suggest that pure dipole transitions are produced only in limited cases. If multiply charged projectiles are utilized, two-center effects are likely to be important. In this case, dipole transitions are expected only for very high projectile energies. Therefore, we consider the collision system Ar^{18+} + Li for an impact energy of 95-MeV/u corresponding to an incident velocity of 59 a.u. [*Stolterfoht* et al. (1991)]. For these high projectile energies, soft-collision electrons have been measured at a limited number of angles which may be compared with theoretical results obtained using the CDW-EIS code by *Gulyás* et al. (1995a). In the following, we shall focus our attention on the theoretical aspect. It is noted that the CDW-EIS data were found to be nearly equal to corresponding results from the Born approximation indicating that two-center effects are negligible for the present collision system.

The Li target atom has the advantage of having two different shells that contain electrons with significantly different binding energies, i.e., 59.3 eV for 1s and 5.5 eV for 2s. Hence, information about shell effects on the dipole transitions may be achieved. Atomic structure effects on the transition probabilities may be significant. Also, it is interesting to verify dipole transitions for loosely bound electrons. In the limiting case of free electrons separated from the nucleus, dipole transitions by photoabsorption become impossible due to the laws of energy and momentum conservation.

In Fig. 5.10 theoretical results for angular distributions of 1 eV electrons, ejected in 95-MeV/u Ar^{18+} + Li collisions are depicted. The calculations were performed using a recent version of the CDW-EIS code [*Gulyás* et al. (1995a)] which has been extended to extract individual multipole contributions. The CDW-EIS results are given separately for the 1s and 2s orbitals in the two parts of the figure. In each part, the monopole and dipole contributions are compared with the full calculation including all multipoles. The cross section maximum at 90° indicates a preferable production of the magnetic quantum numbers $M = \pm 1$. Also, it is seen from Fig. 5.10a that for the 1s orbital the monopole contribution is rather small. Accordingly, the curve attributed to dipole transitions nearly coincide with that associated with all multipoles.

These findings are quite different from the results for the 2s orbital (Fig. 5.10b). At forward and backward angles, the monopole contribution is larger than the dipole contribution and it remains significant for intermediate angles near 90°. Nevertheless, the interference between the monopole and dipole terms is small, but visible. Furthermore, the curve representing the (coherent) sum of the monopole and the dipole contributions is still significantly smaller than the results from the full calculation containing all terms. This shows that for ionization of 2s electrons, multipoles with $l \geq 2$ contribute appreciably to the 1 eV angular distribution.

To obtain more information about the multipole contributions in 95-MeV/u Ar^{18+} + Li collisions, additional CDW-EIS calculations were performed for emission of 10, 30, and 100 eV electrons. The results are shown in

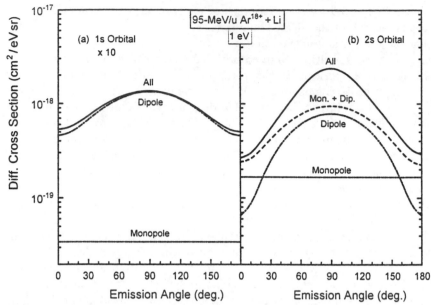

Fig. 5.10a,b. Angular distributions of 1 eV electrons produced in 95-MeV/u Ar^{18+} + Li collisions. In (**a**) calculations are shown for 1s electrons and in (**b**) for 2s electrons. Different multipole contributions are compared with the full calculations including all multipoles. They are obtained using the CDW-EIS code of *Gulyás* et al. (1995a,b). Note that all 1s results have been multiplied by 10

Fig. 5.11 where the coherent sum of monopole and dipole contributions are compared with a full calculation including all multipoles. Parts (a) and (b) show cross sections for the ionization of the 1s and 2s orbitals of Li, respectively. Part (c) gives the sum of the cross sections for the 1s and 2s orbitals in comparison with the experimental data taken at electron emission angles of 25°, 60°, 90°, and 120° [*Stolterfoht* et al. (1991)]. Within the instrumental uncertainties, the experimental data agree well with theoretical results representing all multipoles. This provides confidence that the CDW-EIS does a reliable job in describing the present ionization processes.

Figure 5.11 shows that the characteristic features of the theoretical data become more dramatic as the electron energy increases. The monopole-dipole description is found to be suitable for ionization of the 1s electron even at the highest electron emission energy of 100 eV studied here. Similarly, for 2s ionization the full calculation is reproduced quite well at forward and backward angles. In contrast, when the electron energy increases, at angles near 90° the full calculations are strongly enhanced with respect to the monopole-dipole contributions. The 100 eV data provide evidence that the spectrum involves two components, i.e., a sharp peak at 90° due to binary-encounter collisions and a broad structure representing monopole and dipole transitions. We expect that the broad dipole-monopole structure is associated with the inner

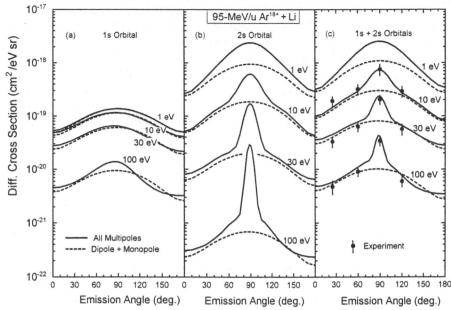

Fig. 5.11a–c. Angular distributions of electrons produced in 95-MeV/u Ar^{18+} + Li collisions evaluated using the CDW-EIS code of *Gulyás* et al. (1995a,b). The electron energies are 1, 10, 30, and 100 eV as indicated at the curves. Each part (**a**), (**b**) and, (**c**) shows the coherent sum of monopole and dipole contributions in comparison with results including all multipoles. In (**a**) data are given for 1s electrons, in (**b**) for 2s electrons, and in (**c**) the sum of the 1s and 2s data in conjunction with experimental from *Stolterfoht* et al. (1991)

part of the 2s wave function (which has the character of a 1s wave function) and the sharp binary-encounter peak is associated with the dilute outer part of the 2s wave function.

The important feature of the present data is that the monopole-dipole contributions for the 1s and 2s orbitals behave quite differently as the electron energy increases. At 1 eV, the 2s contribution is still an order of magnitude larger than the 1s contribution. However, it is remarkable that at 100 eV the monopole-dipole part of the 2s cross section is an order of magnitude smaller than the corresponding 1s cross section. Evidently, the 2s electron is inefficient in receiving an energy as high as 100 eV via a dipole-like transition.

One may be tempted to explain this observation by atomic structure effects of the 2s electron. However, we would expect that the relatively small binding of the 2s electron is primarily responsible for the present observation. A dipole-like transition requires a third body (i.e., the target atom) to accept part of the momentum received in a violent collision. Therefore, a free electron cannot absorb a photon as mentioned earlier. Since the 2s electron is loosely bound, it is difficult to transfer excess momentum to the target and, hence, we expect a small probability for dipole transitions.

Thus, the binary-encounter mechanism, which does not involve a third body, is strongly favored in violent collisions with loosely bound electrons. It is recalled that the production of binary-encounter electrons is accompanied by transfer of high angular momenta. The data in Fig. 5.11b indicate that these binary-encounter electrons are primarily responsible for the discrepancies between the monopole-dipole contribution and the full calculation. It is noteworthy that the binary-encounter mechanism strongly contributes to the production of 1 eV electrons which are usually attributed to soft-collisions.

The present data demonstrate that the predominance of certain multipole terms depends strongly on the shell properties of the bound electron. Dipole transitions are found to be dominant for an electron initially bound in the 1s orbital, whereas ionization of a loosely bound 2s electron implies a broad spectrum of multipoles. It is important to realize that the dipole approximation is generally not valid for 2s ionization by projectiles with energies as high as 95 MeV/u (corresponding to a velocity of 59 a.u.). One may suppose that the 2s ionization by Li represents a rather particular case. However, this finding and the following results for lower projectile energies imply that the dipole approximation should be used with some care when ionization is described in energetic ion-atom collisions.

Low Projectile Energies. In the previous section it was shown that the importance of two-center effects increases with increasing projectile charge state. As two-center effects are governed by the Bohr parameter Z_p/v, we expect an increase of these effects with decreasing projectile energy. Two-center effects produce an angular distribution asymmetric around $90°$ with a clear enhancement at forward angles. The earliest observation of such asymmetric angular distributions for electrons with energies as low as 1 eV was reported by *Rudd* and *Jorgensen* (1963), as also discussed further below. Subsequently, several studies have exhibited asymmetric angular distributions of soft-collision electrons (some of them will be discussed in Sect. 5.3.2). The observed asymmetry is mainly due to the attraction of the outgoing electron by the receding projectile in agreement with the two-center picture discussed previously [*Oldham* (1967), *Stolterfoht* et al. (1987), *Pedersen* et al. (1991)].

Recently, *Suárez* et al. (1993a) have studied electron angular distributions for slow H^+ impact. The results for the collision system 106-keV H^+ +Ne are given in Fig. 5.12 which refers to angular distributions of low-energy electrons ranging from 1 eV to 10 eV. Apart from the asymmetry of the angular distributions, also previously observed, the work by *Suárez* et al. (1993a) revealed important aspects. It has drawn the attention to the analysis of the angular distribution in terms of a multipole expansion. The solid line in Fig. 5.12 results from a fit of the experimental data by the first three terms of the multipole expansion given by (5.6). Further studies have been performed by *Cravero* and *Garibotti* (1993, 1994) using both B1 and CDW-EIS models with hydrogenic wave functions. These analyses are analogous to the investigation of the forward-backward asymmetry of the electron cusp which has

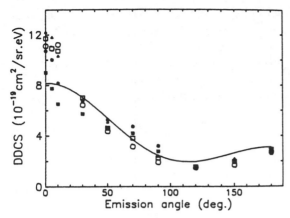

Fig. 5.12. Angular distributions for electrons emitted at a given energy in collisions of 106-keV H$^+$ on Ne. Different electron energies are represented by different
symbols: *Solid square:* 1 eV; *solid circle:* 2 eV; *solid triangle:* 5 eV; *open square:* 7 eV
and *open circle:* 10 eV. The solid line results from a fit of the experimental data
by means of a multipole expansion including the first 3 terms. From *Suárez* et al.
(1993a)

received much attention in the past [*Meckbach* et al. (1981), *Burgdörfer* et
al. (1984)].

The analysis of the soft collision electrons by *Suárez* et al. (1993a) left
some open questions. Following the studies by *Briggs* and *Day* (1980), it was
argued for the case of a hydrogenic target that the B1 approximation predicts symmetric angular distributions at asymptotically small energies and,
hence, the observed asymmetry was attributed uniquely to two-center effects. However, the question of whether the hydrogenic case can be extended
to multielectron targets remained open. This question has recently been addressed by *Fainstein* et al. (1996) showing that hydrogenic and multielectron
targets yield different results.

In Fig. 5.13 we show data for angular distributions of soft-collision electrons produced by about 100 keV H$^+$ impact on H and He. The experimental
results are taken from the work by *Kerby* et al. (1995) and *Rudd* and *Jorgensen* (1963), respectively. The theoretical results are obtained by means of
the CDW-EIS and B1 approximation using the code by *Gulyás* et al. (1995a)
as mentioned before. For 1 eV electrons a comparison between experiment and
theory is possible. The CDW-EIS calculation is in good agreement with the
experiment supporting two-center effects as the main reason for the observed
asymmetry.

Inspection of Fig. 5.13b shows that the Born approximation does not yield
a symmetric angular distribution. Evidently, the asymptotic limit for low-
energies is not yet reached at 1 eV. To verify the CDW-EIS and B1 approximation at the asymptotic energy limit, cross sections for the emission of
10^{-4} eV electrons were calculated (Fig. 5.13a). Indeed, the B1 data indicate

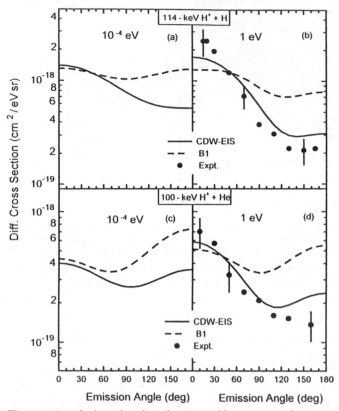

Fig. 5.13a–d. Angular distributions of low-energy electrons produced in collisions of 114-keV H^+ on atomic hydrogen and 100-keV H^+ on helium. The experimental data for 1 eV shown in (**b**) are from *Kerby* et al. (1995) and (**d**) from *Rudd* and *Jorgensen* (1963). The theoretical results are obtained from the CDW-EIS and B1 approximation using the program by *Gulyás* et al. (1995a). Corresponding theoretical data for 10^{-4} eV electrons are shown in (**a**) and (**c**)

a symmetrical angular distribution at the asymptotic limit that has also been predicted by means of the semi-classical approximation in Sect. 3.3.3.

It is important to realize that for angular distributions the asymptotic limit is reached at 10^{-4} eV but not at 1 eV. The estimate of the single-differential cross sections by means of (3.5) has shown that the asymptotic value is practically attained at 1 eV. It was noted that the single-differential cross sections are governed by the related energy transfer that does not change much at the low electron energy limit (as this transfer includes the binding energy).

However, from (3.71) it follows that the angular distribution is influenced by the phase of the continuum wave function. This phase depends critically on the potential of the target center and it may still change appreciably below 1 eV (or even below 0.1 eV). Accordingly, the angular distribution of the soft-collision electrons may undergo a significant variation as the electron energy tends to zero.

The asymptotic behavior becomes even more complex when other target atoms are used. From Fig. 5.13c it is evident that the finding for the hydrogenic target cannot be generalized to other target atoms. For He the Born approximation does not yield a symmetrical angular distribution at asymptotically small energies of 10^{-4} eV. It is interesting to note that such variation has not as yet been confirmed experimentally; see also the threshold results in the compilation by *Rudd* et al. (1976).

Hence, an asymmetric angular distribution observed at 1 eV for non-hydrogenic targets may have different origins. Besides two-center effects, the formation of the target center alone produces asymmetries. Nevertheless, the comparison of B1 and CDW-EIS results indicates that two-center effects likely are most important for proton energies as low as 100 keV. To obtain more information about the theoretical one- and two-center calculations it is useful to perform multipole expansions. In particular, further work is needed to analyze the angular distributions of soft-collision electrons at a more complete set of projectile energies.

5.2.4 Backscattering of High-Energy Electrons by the Target

The backscattering in the target center is also important for electrons that have suffered more than just a soft collision with the projectile. In the violent collisions considered here, typical energies transferred to the ejected electrons amount to a few hundred eV. In this case, the probability for subsequent electron backscattering is small. Nevertheless, since other mechanisms, such as the inverse binary-encounter process (Fig. 2.3b), have even smaller probabilities, backscattering by the target dominates. In fact, it will be shown below that the backscattering may enhance the corresponding cross sections for electron emission by more than an order of magnitude.

Recalling Fig. 5.7, large discrepancies occur between the experimental and theoretical data at backward angles. These discrepancies originate from effects of the target center. *Rudd* et al. (1966a) used an early version of the Born approximation that is based on scaled hydrogenic wave functions. This approximation implies an effective nuclear charge of the target that is smaller than the corresponding bare nuclear charge. Backscattered electrons result from collisions close to the nucleus where the electron is subject to the full target nuclear charge rather than a smaller effective charge which is appropriate for greater distances. Therefore, the strong electron-target backscattering is underestimated by these calculations based on the scaled hydrogenic wave functions.

Figure 5.7 indicates that the Born approximation underestimates the experimental data at backward angles more strongly than expected from the reduction of the target charge adopted in the hydrogenic approach. Hence, additional effects should play a role. As shown in Fig. 5.1, the target backscat-

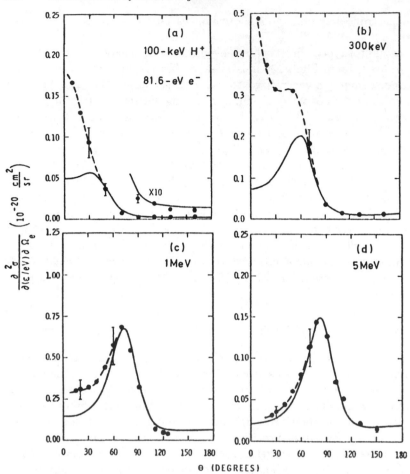

Fig. 5.14a–d. Angular distributions of 81.6 eV electrons ejected in H$^+$ + He collisions at different projectile energies. The experimental results of 100 keV and 300 keV are from *Rudd* et al. (1966a). The 1-MeV and 5-MeV data are from laboratories in Richland and Berlin included in the compilation by *Rudd* et al. (1966a). The data are given by *Manson* et al. (1975) who provided also the theoretical data. The cross section compilation is made by *Stolterfoht* (1978)

tering resembles the production of binary-encounter electrons at forward angles implying a backscattering of the electrons in the projectile center. It has been found that large-angle scattering is significantly enhanced in a non-Coulombic field [*Richard* et al. (1990)]. Similar effects are likely to determine the enhanced electron ejection at backward angles. Large angle scattering will be discussed in detail later, when the binary-encounter peak is treated for dressed projectiles.

The importance of an adequate description of the final continuum state was recognized by *Madison* (1973) who used Hartree-Fock-Slater wave functions in the Born approximation. Similar calculations were made by *Manson* et al. (1975). With the inclusion of realistic bound and continuum wave functions, a crucial source for discrepancies between theory and experiment was eliminated. This is shown in Fig. 5.14 where theoretical results by *Manson* et al. (1975) are displayed in comparison with experimental cross sections for emissions of 81.6 eV electrons by H^+ impact with energies ranging from 100-keV to 5 MeV. The experimental data, given also in tabulated form by *Rudd* et al. (1976), are from research groups at the University of Nebraska, the Pacific Northwest Laboratories, and the Hahn-Meitner-Institut Berlin. As shown in the figure, the theoretical results are found to be in remarkable agreement with the experimental cross sections for ejection at backward angles. This excellent agreement indicates that the Born approximation, when used with adequate Hartree-Fock-Slater wave functions in both initial and final states, is suitable for accurately predicting the data at backward angles.

However, Fig. 5.14 shows that significant deviations between theory and experiment remain at forward angles. These discrepancies, attributed to two-center effects, are seen to diminish when the projectile energy increases. It is important to note that for the relatively high projectile energy of 5 MeV, the theory agrees very well with the experiment at all angles (Fig. 5.14d). This provides confidence that the Born approximation can be used as a standard to verify two-center effects. This will be done in next sections.

5.3 Two-Center Phenomena

Two-center effects are important when the ejected electrons are significantly influenced by the fields of both the projectile and target (Fig. 2.2). In the electron spectra, two-center phenomena are observed in regions that are not dominated by one-center effects. As pointed out in the previous section, two-center effects are identified experimentally by comparing measured cross sections with the corresponding model results of the Born approximation which represents the contribution from a single center. Alternatively, two-center effects may be identified exclusively from experiment by comparing results using high incident charge states with equal velocity data for proton impact. However, this method works only for energies above a few MeV/u, since low-energy protons also give rise to two-center effects (Fig. 5.13).

Two-center effects have been observed since the beginning of electron spectroscopy experiments. Early observations of enhanced electron emission at forward angles by *Rudd* and *Jorgensen* (1963) motivated *Crooks* and *Rudd* (1970) to study electron emission at an angle of 0°. Hence, they discovered a pronounced cusp-shaped peak which they attributed to the process of electron transfer to the continuum. More recently, various studies of two-center effects have been performed [*Fainstein* et al. (1991), and references therein]. Some of these studies will be discussed in the following sections.

To classify two-center effects it appears useful to consider the time correspondence of the interactions of the two centers during the collision. The concept of two-center effects implies a certain coexistence of the associated nuclear fields. Nevertheless, usually there are sequential actions by the two centers. For instance, the ECC electrons are initially bound to the target and later feel a dominant projectile field. On the other hand, there are cases, such as the "saddle-point electrons", where the simultaneous action of the two centers is an important aspect of the interaction. In the following sections, two-center effects are discussed in terms of the simultaneity of the two-center actions. After a brief treatment of the ECC mechanism, saddle-point electrons, and two-center electron emission are reviewed.

5.3.1 Electron Capture to the Continuum

The process of electron capture to the continuum, producing the electron cusp, has received a great deal of attention during the past decades. The studies concerned with this process are so numerous that a complete review cannot be accomplished within the scope of the present work. Therefore, we shall limit the presentation of the experimental results to a few examples showing the fundamental aspects of the ECC process. Rather, the emphasis is focused on the connections with other two-center effects. The reader who is interested in more details is referred to the articles by *Breinig* et al. (1982), *Köver* et al. (1983), *Burgdörfer* (1984), *Závodszky* et al. (1994), *Pregliasco* et al. (1994), *Menendez* et al. (1991), and *Berényi* et al. (1994).

The first indications for cusp electrons were observed at forward angles in the pioneering experiments by Rudd et al. (1966a), who studied double-differential cross sections for proton impact at intermediate energies. The data for 100 keV protons measured at an electron observation angle of 10° exhibited an unexpected hump at about 50 eV which corresponds to an electron velocity approximately equal to the projectile velocity. This structure was associated with the interaction of electrons traveling close to the projectile after the collision [*Oldham* (1967), *Salin* (1969), *Macek* (1970)].

To verify the idea of the electrons traveling with the projectile, *Crooks* and *Rudd* (1970) performed experiments at the observation angle of 0° expecting a significant enhancement of the electrons ejected with a velocity equal to the projectile velocity. The result of their measurement was indeed remarkable. As shown in Fig. 5.15, the double differential cross sections for electron emission at 0° exhibit a cusp-shaped peak that exceeded the background by nearly an order of magnitude. A similar observation was made by *Harrison* and *Lucas* (1970) who observed electron emission from protons passing through a foil. To interpret this cusp, the "post-collisional" interaction of the projectile-electron system was interpreted as a mechanism of charge-transfer to the projectile continuum [*Crooks* and *Rudd* (1970), *Macek* 1970), *Dettman* et al. (1974)].

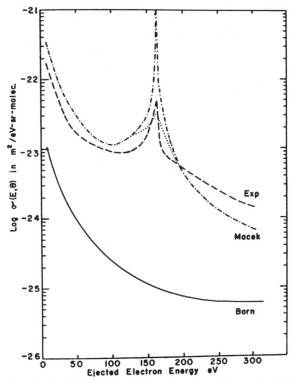

Fig. 5.15. Electron cusp produced at forward angles by 300-keV H^+ impact on He. The *dashed* and *dotted curves* refer to experimental data observed at $0°$ and $1.4°$. The *solid* and *dashed dotted curves* represent calculations using the Born approximation and the theory by *Macek* (1970). The *dotted line* is due to the model by *Salin* (1968). From *Crooks* and *Rudd* (1970)

The theoretical treatments by *Salin* (1969) and *Macek* (1970) clearly show that the ECC mechanism requires a description beyond the target-centered Born approximation. The ECC implies two-center phenomena, as long-range forces of both the projectile and target atoms play an important role in the formation of the electron cusp. In Fig. 5.15 the experimental data are compared with results obtained from the Born approximation and using the theoretical approach by *Macek* (1970). The Born approximation fails to reproduce an enhancement of the cross sections near the cusp. However, the theory by *Macek* (1970), which implies higher-order terms of the Born series (Sect. 3.5.4), is able to account for the experimental observation. Similar results have been obtained by *Salin* (1972).

In Fig. 5.16 more recent data from calculations of the cusp by *Schultz* and *Reinhold* (1994) are compared to experimental results by *Lee* et al. (1990) for the collision system $1.5\,\mathrm{MeV/u}\ F^{9+} + He$. As expected, the B1 approximation (dotted line) does not account for the cusp electrons. However, the cusp is

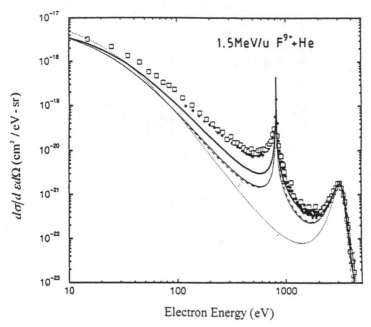

Fig. 5.16. Spectrum of electrons ejected at 0° in the laboratory for collisions of 1.5 MeV/u F^{9+} with helium. *Solid circles:* experimental data of *Lee* et al..(1990) and the theoretical results are from *Schultz* and *Reinhold* (1994), *open squares:* CTMC result, *solid line:* CDW-EIS calculations using effective Coulomb target fields, *heavy solid line:* model potential-CDW-EIS calculation; *dotted line:* B1 results, and *dashed line:* DSPB predictions [*Brauner* and *Macek* (1992)]. From *Schultz* and *Reinhold* (1994)

reproduced by two-center theories such as the CTMC [*Schultz* and *Reinhold* (1994)], the CDW-EIS [*Schultz* and *Reinhold* (1994)], and the DSPB [*Brauner* and *Macek* (1992)]. The CTMC with the model potential by *Garvey* et al. (1975) is found to be in favorable agreement with the experiment.

A plausible scheme to explain the production of the cusp electrons is the classical double-scattering mechanism by *Thomas* (1927). From two-body kinematics it follows that an electron with a velocity equal to the projectile velocity is produced in a binary-encounter collision at an angle of 60°. As seen from Fig. 2.3f this single interaction would lead to an electron rapidly separating from the projectile so that the process of electron capture becomes impossible. However, if it undergoes a second scattering by 60° in the field of the target nucleus, the electron may travel on a trajectory parallel to the projectile direction corresponding to capture to the projectile continuum.

The Thomas mechanism is not expected to contribute strongly to electron capture to the continuum [*Dettmann* et al. (1974)]. Nevertheless, the classical double scattering picture is used here for a qualitative discussion of the ECC. The Thomas picture shows that both target and projectile play an important

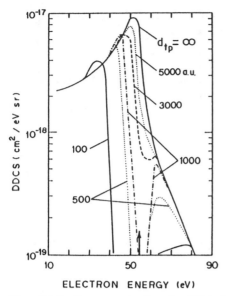

ELECTRON ENERGY (eV)

Fig. 5.17. Cusp production in 100-keV H^++ He collisions calculated using the CTMC method. The data are evaluated with a Hamiltonian whose integration is stopped at different internuclear distances d_{tp} as indicated. From *Reinhold* and *Olson* (1989)

role for the cusp formation and that the projectile interacts with the electron for a long time.

It is instructive to view the cusp production within the framework of classical mechanics. The double scattering mechanism displayed in Fig. 2.3f suggests that the electrons are focused in the forward direction. This focusing takes place on a time scale large in comparison with that of the collision [*Swenson* et al. (1989)]. If the focusing is directed to a point at infinity, the electron intensity exhibits a singularity at 0°. Hence, the cusp profile characteristic for the ECC process may be produced. It should be emphasized, however, that there is no great difference between electron capture to the continuum and Coulomb focusing. In fact, the focusing phenomenon is the classical counterpart to the ECC mechanism.

The picture of electron focusing is supported by classical-trajectory Monte-Carlo calculations [*Reinhold* and *Olson* (1989), *Montemayor* and *Schiwietz* (1989)] and by related impact-parameter measurements [*Jagutzki* et al. (1991b)]. To demonstrate the focusing aspect in the cusp formation, *Reinhold* and *Olson* (1989) terminated their calculations before the collision partners have separated completely. Figure 5.17 shows corresponding CTMC results for the cusp-peak formation in 100-keV H^+ + He collisions where the integration of the Hamiltonian is terminated a long time after the collision. Nevertheless, internuclear distances of more than several thousand a.u. are needed to achieve the characteristic cusp observed for the ECC process.

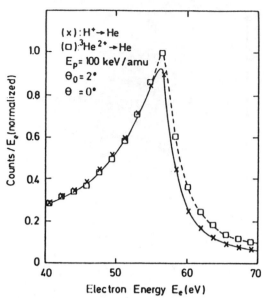

Fig. 5.18. Double differential cross sections for electron emission by 100-keV/u He^{2+} and H^+ impact on He. Asymmetric ECC cusps are shown. The observation angle is 0°. The H^+ data are multiplied by a factor of 4. From *Bernardi* et al. (1989)

The present consideration shows that the electron-projectile interaction at large internuclear distances plays an essential role in the formation of the cusp electrons [*Reinhold* and *Olson* (1989)]. However, it should be kept in mind that the classical picture has its limitations. From post-collisional focusing it is expected that the electron intensity is redistributed in the angular space, whereas the integrated cross section remain unchanged. However, detailed studies of the electron cusp have shown that the ECC mechanism involves also an enhancement of the cross section suggesting a mechanism which exhibits characteristic features of a capture process.

Despite the importance of the long-lasting electron projectile interaction, the role of the target field should not be underestimated. The experimental data show that the cusp profile is asymmetric. The asymmetry of the cusp profile has been revealed by comparison with the multipole-expansion (5.6) as has been done in several laboratories [e.g., *Meckbach* et al. (1981), *Burgdörfer* (1984, 1986), *Berry* et al. (1985)].

Examples for the asymmetry of the cusp are given in Fig. 5.18 where results for 100-keV/u H^+ and He^{2+} impact on He are plotted [*Bernardi* et al. (1989)]. The observed asymmetries are explained by the attraction of the captured electron to the receding target ion. In the projectile frame of reference, the target field drags the electrons to the backward direction which, in turn, causes an enhancement of the low-energy side of the cusp observed in the laboratory frame. These considerations show that the projectile and

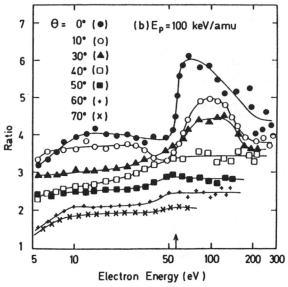

Fig. 5.19. Double differential ratio $R(E_e, \theta)$ of electron emission induced by 100-keV/u He^{2+} and H^+ impact on He. Parameter indicated at the *curves* is the observation angle. The results for 0° are based on the data in Fig. 5.18. The *arrow* indicates the ECC peak energy where the projectile velocity equals the electron velocity. From *Bernardi* et al. (1989)

target fields simultaneously act on the outgoing electron. Hence, as the target field plays a noticeable role even at large internuclear distances, the electron cusp should not be regarded in a simplified post-collision picture.

Another way of looking at the asymmetry of the ECC peak is the analysis of cross section ratios. In Fig. 5.19 the cross sections for electron emission by 100-keV/u He^{2+} impact are displayed as a ratio to the corresponding results for 100-keV H^+ [*Bernardi* et al. (1989)]. The curves refer to different electron observation angles including 0°. The cross section ratios are rather constant until about 27 eV. At this energy the 0° curve exhibits a sudden rise by more than a factor of 2. Weaker increases are seen for the forward angles of 10° and 20° whereas the cross section ratios remain essentially constant for the other angles. The sudden rise occurs at the cusp maximum where the projectile velocity equals that of the electron (see the arrow at 27 eV).

The sudden increase of the 0° curve is explained by the asymmetry of the underlying cusp. The asymmetry is governed by the field of the receding target ion (He^+ in this example) in relation to the field of the projectile. Hence, the relative strength of the target field increases with decreasing charge of the projectile and the cusp asymmetry is larger for H^+ than for He^{2+}. The sudden rise of the cross section ratio is due to the fact that the cross sections of a more symmetric cusp (for He^{2+}) are divided by results of a more asymmetric peak (for H^+). The finding that the cross section ratio is approximately

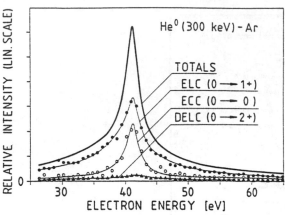

Fig. 5.20. Electron cusp for 300-keV He^0 + Ar collisions measured in coincidence with the scattered projectiles. The projectile charge state 0, 1, and 2 refer to electron capture to continuum (ECC), electron loss to continuum (ELC) and double electron loss to continuum (DELC), respectively. From *Sarkadi* et al. (1989)

constant on the left cusp wing indicates that the underlying cusps are quite similar at the low-energy side. The difference between the He^{2+} and H^+ cusps occurs only at the high energy side [*Bernardi* et al. (1989)]. This phenomenon has also been observed in CTMC calculations [*Reinhold* and *Olson* (1989)]. However, it appears that a satisfying explanation is still missing.

Another observation that attracted much controversial attention is the cusp production by neutral projectiles. *Sarkadi* et al. (1989) performed demanding experiments where the charge states of the scattered particles are measured in coincidence with the cusp electrons. Incident He^+ ions were first used and the cusp profile was measured in coincidence with neutral He, i.e., with scattered projectiles that have captured an electron into bound states in addition to the capture of an electron into continuum states. In this case, the process producing the cusp refers to a transfer-ionization mechanism [*Andersen* et al. (1984)]. The experimental data showed a surprisingly narrow cusp which motivated *Sarkadi* et al. (1989) to use neutral He as incident particles.

The results are plotted in Fig. 5.20 which provides clear evidence for a cusp-shaped electron peak produced by neutral ingoing and outgoing particles. Close inspection of the figure indicates that the cusp attributed to neutral particles is narrower than that for charged particles. This observation is a major challenge for theoretical interpretations. *Garibotti* and *Barachina* (1983) have shown that potentials, dropping to zero faster than the Coulomb potential, can produce a cusp profile. Another promising explanation refers to the capture of electrons into excited projectiles forming rapidly decaying resonances. Recently, *Sarkadi* et al. (1997) considered low-lying resonance states of He^- formed by capture into the states 2^1S and 2^3S of metastable He beam components.

In view of the cusp formation, interesting questions remain open. For example, when the double scattering mechanisms by Thomas does not provide a strong contribution to ECC, other processes are to be considered [*Dettmann* et al. (1974)]. Capture to the continuum may proceed via a single-step event incorporated in the first-order theory by *Brinkmann* and *Kramers* (1930). This mechanism involves a weak interaction with the target and, hence, suggests that ECC corresponds predominantly to a single-center phenomenon associated with the projectile [*Ovchinnikov* and *Khrebtukov* (1987)]. It should be realized, however, that a projectile single center treatment such as the binary-encounter theory does not reproduce the cusp. Hence, at least a first-order interaction with the target atom is required during the collision.

Moreover, if two centers are adopted, it is interesting to verify whether their actions are independent events occurring successively. It is recalled that the cusp is generally observed to be asymmetric [*Berry* et al. (1985)], an effect which is commonly attributed to the influence of the receding target atom. This shows that the influence of the target atom cannot be neglected in the post-collision region. Hence, although we expect the projectile field to be dominant at large internuclear distances, the cusp electrons are simultaneously influenced by the field of both collision partners.

5.3.2 Two-Center Electron Emission

Two-center electron emission may be understood as a generalization of the ECC process. The TCEE mechanism involves a deflection without focusing of the outgoing electron in the Coulomb field of the receding projectile, see Fig. 2.3e. Two-center effects are particularly important for electrons ejected at forward angles including 0° [*Meckbach* et al. (1986), *Bernardi* et al. (1989)]. However, TCEE implies that electrons, emitted at backward angles, may also feel the projectile field in addition to that of the target atom. Thus, TCEE has been observed in the full angular range of electron emission [*Stolterfoht* et al. (1987), *Schneider* et al. (1989), *Pedersen* et al. (1990)]. In Sect. 5.2.3, two-center effects have been treated for soft collisions. Here, it will be shown that two-center effects gain importance for electrons ejected with increasing energy. Two-center effects have previously been reviewed by *Fainstein* et al. (1991) so that the following discussion shall be limited to specific examples.

It is most advantageous to carry out the TCEE measurements with fast ions, i.e., for which the projectile velocity v is much larger than the orbital velocity v_e of the active electron. This is due to the fact that the spectral structures originating from one-center effects become more and more separated as the projectile energy increases. Looking back at Fig. 2.1 it is recalled that TCEE is expected in the region between the SC and ECC peak. This range becomes larger when high-energy projectiles are used. It should be realized, however, that the different electron production mechanisms are not distinguishable. Therefore, for a detailed study of two-center electron emis-

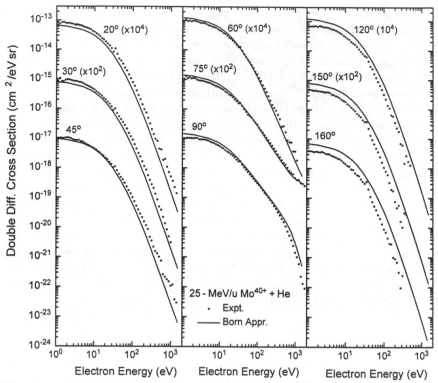

Fig. 5.21. Double-differential cross sections for electron emission in 25-MeV/u Mo^{40+} + He collisions for different observation angles as a function of the electron energy. The theoretical results are obtained using the Born approximation. From *Stolterfoht* et al. (1987). A few revisions were made with respect to the soft-collision electrons [*Stolterfoht* et al. (1995)]

sion, high projectile energies are useful as they provide a spectral region where two-center effects are dominant.

Results for high-energy collisions are shown in Fig. 5.21 where cross sections for electron emission by 25-MeV/u Mo^{40+} impact on He are plotted for different observation angles [*Stolterfoht* et al. (1987, 1995)]. The plot displays the experiment in comparison with theoretical results that are obtained by means of the Born approximation with Hartree-Fock-Slater wave functions for the initial and final continuum state. The use of accurate wave functions is essential, as the Born approximation is applied here as a standard of comparison to verify the appearance of two-center effects. Since the Born approximation accounts uniquely for target-center effects, it can be presumed that discrepancies between experiment and theory provide evidence for two-center phenomena.

Figure 5.21 shows that for angles around 90° experiment and theory agree well above ∼ 100 eV. At these angles and energy the dominant part of the electron emission is due to the binary-encounter process. Note for instance the broad shoulder near 2 keV in the 75° spectrum. It is recalled from (2.1) that the binary-encounter maximum is predicted for electron emission at angles smaller than 90°. However, binary-encounter electrons may also appear at angles somewhat larger than 90° as the electron is scattered in the field of the target nucleus. This effect is incorporated in the Born approximation which accounts for all interactions with the target nucleus. Therefore, the good agreement between theory and experiment is understood for angles around 90°. However, at angles much different from 90°, the experimental cross sections deviate significantly from the predictions of the Born approximation. From Fig. 5.21 it is seen that the experimental cross sections are enhanced at forward angles (e.g., 20°) and reduced at backward angles (e.g., 160°) in comparison with the B1 results.

The observed enhancement at forward angles and the reduction at backward angles can be associated with the occurrence of two-center effects [*Stolterfoht* et al. (1987)]. This supposition is supported by CDW-EIS calculations which includes both centers in the final state [*Fainstein* et al. (1988a)]. It should be noted, however, that previous studies have suffered from inaccuracies of the CDW-EIS calculations which resulted from using hydrogen-like wave functions for the ejected electrons. As pointed out before, this deficiency of the CDW-EIS theory has recently been removed. As in the work by *Madison* (1973) and *Manson* et al. (1975), Hartree-Fock-Slater wave functions were implemented to describe the final continuum state centered at the target atom [*Fainstein* et al. (1994), *Gulyás* et al. (1995a,b)]. Thus, more recently, a sensitive analysis of the experimental data has become possible.

For a detailed comparison of the experimental and theoretical results, it is advantageous to consider angular distributions rather than energy distributions [*Manson* et al. (1975)]. As seen from Fig. 5.21, the double differential cross sections change by orders of magnitude in the measured electron energy range. In this case, discrepancies between theory and experiment are often lost in the graphical display of the cross sections. However, the cross sections vary less with the electron emission angle so that a sensitive graphical comparison is possible in this case.

Figure 5.22 shows a comparison between experimental and theoretical cross sections as a function of electron emission angle for Ne^{10+} impact on He [*Stolterfoht* et al. (1995)]. The experimental results are revised data from *Platten* et al., (1987) and the theoretical results are obtained using the B1 and CDW-EIS code by *Gulyás* et al. (1995a). For all electron energies the Born approximation underestimates the experimental data at forward angles and overestimates them at backward angles. In contrast to the B1, the CDW-EIS shows excellent agreement with the experiment in the complete angular range and at all energies. It is important to keep in mind that electron energies are chosen within a range covering three orders of magnitude (Fig. 5.22). Such an agreement between absolute cross sections in nearly complete regions of both the electron emission angle and energy is remarkable. It shows that the

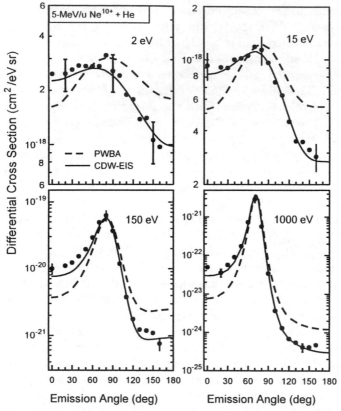

Fig. 5.22. Double-differential cross sections for electron emission in 5-MeV/u Ne[10+] + He collision as a function of the electron observation angle. A few electron energies are selected as indicated. The experimental results are compared with calculations using the B1 and CDW-EIS. From *Stolterfoht* et al. (1995)

CDW-EIS theory with Hartree-Fock-Slater wave functions does an excellent job in describing the specific features of two-center effects occurring at high projectile velocities.

At relatively high electron energies of 150 eV and 1000 eV, Fig. 5.22 shows that two-center effects are mostly noticeable in regions where the cross sections are small. Other regions, where the cross sections are larger, such as at the binary encounter peaks near 75° (Fig. 5.22), one-center phenomena dominate and therefore the Born approximation agrees well with experiment. In fact, the best agreement between B1 and experiment is observed for 150 eV and 1000 eV. At lower energies (15 eV) the binary-encounter peak is shifted with respect to the B1 results. This angular shift is analogous to the energy shift of the binary-encounter peak which has previously been studied in terms of two-center effects (Sect. 5.1.2).

Figure 5.22 shows an enhancement of the angular shift for the 2-eV electron maximum which, in turn, produces an asymmetry of the angular distribution with respect to 90°. This asymmetry has been studied for the present collision systems by *Colavecchia* et al. (1995). A similar collision system 3.6 MeV/u Ni^{24+} + He, investigated via coincidences with recoil ions by *Moshammer* et al. (1994), exhibited asymmetries for electron energies as low as 0.5 eV. Hence, two-center effects cannot be ignored for soft-collision electrons as discussed before in Sect. 5.2.3. It should be realized, however, that two-center effects are relatively weak for soft-collision electrons. In the present collision systems, they do not exceed a factor of two for 2 eV electrons whereas they are as large as a factor of five for 1000 eV electrons (Fig. 5.22).

The two-center effects produced by highly charged projectiles involve specific features (Fig. 2.3e) that may be interpreted in terms of a post-collision picture by a deflection of the electron in the field of the projectile nucleus after leaving the target center [*Stolterfoht* et al. (1987), *Schneider* et al. (1989), *Pedersen* et al. (1990)]. These two-center effects enhance the electron intensity at forward angles. Likewise, the electron intensity is reduced for electrons ejected at backward angles. Obviously, for a projectile with a high charge, the ejected electron is significantly attracted even when it moves rapidly in a direction opposite to that of the projectile.

However, as for the ECC mechanism, the post-collision picture should not be overestimated, i.e., one should not overlook the role of the target. Recalling Fig. 2.3e it is evident that the deflection of the electron in the target field is an important part of the electron emission process. Hence, it should be kept in mind that the target nucleus plays an essential role for the two-center effects in fast highly-charged collision systems. For each particular case it remains to be shown whether the target and projectile centers act simultaneously or successively.

More information about the present type of two-center effects can be achieved when cross section ratios rather than absolute values are compared. Figure 5.23 shows experimental and theoretical cross sections divided by the corresponding data from the Born approximation [*Stolterfoht* et al. (1995)]. The plot is based on the experimental data given in Fig. 5.22. Apart from the redistribution of the electron intensity from backward to forward angles, two-center effects by high-energy projectiles reveal further specific features.

First, the figure confirms the previous finding that two-center effects increase with increasing electron energy. This observation may be understood from the fact that a fast electron feels the two-center field of the rapidly disintegrating collision system more strongly than an electron which emerges slowly from the target. Second, the reduction of the electron intensity at backward angles is larger than the enhancement at forward angles (compared to the Born approximation). It is interesting to note that this behavior is reproduced by both the experimental and theoretical results. The enhanced reduction at backward angles was interpreted by Fainstein et al. (1991) as being due to saturation effects that will be discussed in the following section.

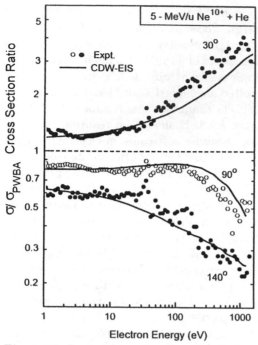

Fig. 5.23. Ratio of cross sections for electron emission in 5-MeV/u Ne^{10+} + He collisions. The data from experiment and CDW-EIS are displayed in relation to the corresponding results from the Born approximation (PWBA). Observation angles are 30°, 90°, and 140° as indicated at the *curves*. The peak structures at 35 eV are due to autoionization electrons from He. The figure is taken from *Stolterfoht* et al. (1995) and is based on the data given in Fig. 5.22

The TCEE data shown in Figs. 5.21,22 were obtained for electron velocities much lower than the corresponding projectile velocity. Hence, further TCEE studies were suggested to extend the electron emission energy to the ECC peak region [*Fainstein* et al. (1989b)]. To cover this region *Pedersen* et al. (1990, 1991) performed measurements of electron emission with comparatively low projectile energies. Cross section ratios for C^{6+} and H^+ projectiles are shown in Fig. 5.24 for impact energies of 1 MeV/u and 1.84 MeV/u. The corresponding ECC peaks are located at 544 eV and 1001 eV respectively. Below these ECC peak energies, the cross section ratios for forward angles increase with increasing electron energy, and at backward angles they decrease with increasing electron energy. These findings are in accordance with the previous results shown in Fig. 5.22.

Figure 5.24 shows that the 20° data increase with energy up to the ECC peak. After passing the ECC peak region, the cross section ratio decreases rapidly. As noted by *Fainstein* et al. (1991) the 20° data transiently reaches unity at the location of the binary-encounter peak. This is expected as the BE peak refers to a one-center mechanism that is well described by the Born

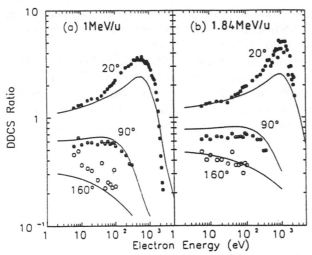

Fig. 5.24a,b. Ratio of cross sections for electron emission from He in collisions with C^{6+} and H^+ as a function of the electron energy. The cross sections for C^{6+} are divided by $Z_p^2 = 36$. In (**a**) results are given for 1-MeV/u impact and in (**b**) for 1.84-MeV/u impact. Electron observation angles are $20°$, $90°$, and $160°$ as indicated. The *points* refer to the experiments by *Pedersen* et al. (1990, 1991) and the *solid curves* refer to the CDW-EIS calculations by *Fainstein* et al. (1991) from which the figure is taken

approximation. Similar effects have previously been noted for electron emission near $90°$ [*Stolterfoht* et al. (1987, 1995)]. In Fig. 5.23 it is seen that the cross section ratio comes closer to unity at a few hundred eV where binary-encounter electrons are expected due to the backscattering in the target center.

From Fig. 5.24 the cross section ratio is seen to drop below unity as the electron energy increases beyond the BE peak. This behavior may be explained by "binding" mechanisms where the outgoing electrons are significantly affected by the attractive force of both collision partners. *Fainstein* et al. (1991) have shown by means of the final CDW wave functions that for asymptotically high electron velocities the nuclear field of the collision partners corresponds to that of the united atom with the nuclear charge of $Z_p + Z_t$. It is plausible that the combined nuclear field attracts the outgoing electron so that the emission cross section is reduced with increasing projectile charge, regardless of the observation angle. This binding phenomenon cannot be accounted for by the Born approximation that describes the emergence of the active electron from a single center formed by the target nucleus only.

Previously, an attempt was made to exploit the importance of the combined nuclear charge for high-velocity electrons in a series of measurements using very heavy collision partners [*Güttner* et al. (1982), and references therein]. It was anticipated that high-energy electrons can be used as a probe to achieve information about super-heavy nuclei transiently formed during the collision. The hope of finding super-heavy elements with the help of electron spectroscopy was not fulfilled, however.

It should also be recalled that two-center effects have been studied in detail at intermediate energies of a few hundred keV [*Bernardi* et al. (1988, 1989), *Suárez* et al. (1993b)]. The studies have been focused on the electron energy and angular ranges where the ECC mechanism is relevant. The corresponding results have been discussed in the previous Section devoted to the cusp electrons. The most important finding is the enhancement of the electron intensity at forward angles. Nevertheless, characteristic differences remain between the studies at low and high impact energies. It should be realized that with increasing electron energy the behavior of the cross section ratios in Figs. 5.19,24 is essentially different. In this case, it is useful to perform further studies verifying two-center effects in the region between low and high impact energies.

In several cases, the analysis of two-center effects has been based on the comparison with the CDW-EIS theory. At high impact energies the CDW-EIS is found to compare very well with experiment, in particular, for 5-MeV/u Ne^{10+} impact (Fig. 5.22). In view of this remarkable agreement between theory and experiment, it is interesting to search for remaining discrepancies. Such deviations are expected since the theory accounts for two-center effects in a perturbative approach. Although differences between experiment and theory are small for 5-MeV/u Ne^{10+}, discrepancies were detected at higher electron energies (Fig. 5.23). The deviations between theory and experiment increase noticeably at forward angles as the electron energy approaches the ECC region (Fig. 5.24). Similar discrepancies have been observed previously for strong interactions in the final state [*Gulyás* et al. (1995a)]. It appears for high incident charge states or low projectile energies that post-collisional effects of the projectile are underestimated by the CDW-EIS approximation. Further work is required to study higher-order terms for two-center effects with projectiles of increasing charge state.

5.3.3 Saddle Point Electron Emission

Another example of two-center electron emission is commonly referred to as "saddle point electron emission". This is a process where both the projectile and target charge centers play strong roles in determining the electron energies and directions. As shown in Fig. 5.25 the potential shape resembles a saddle. It has a maximum along the internuclear axis and a minimum along the direction perpendicular to it. In saddle point ionization, it is predicted that a significant number of electrons can be stranded in the saddle potential which would lead to an enhanced emission of electrons in the forward direction with velocities approximately half that of the projectile. During the past decade, numerous discussions regarding this mechanism have taken place.

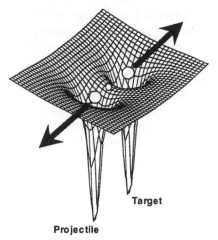

Target

Projectile

Fig. 5.25. Potential wells and saddle existing between receding projectile and target ion. The two larger spheres indicate the respective nuclei while the smaller sphere indicates an electron which is "stranded" on the potential saddle between the two charge centers

At low impact energies this mechanism is expected to be important because other ionization mechanisms are quite weak [*Winter* and *Lin* (1984), *Pieksma* and *Ovchinnikov* (1994), *Pieksma* et al. (1994)]. However, in the 1980s *Olson* (1983, 1986) suggested that this mechanism may play an important role at much higher impact energies, e.g., tens to hundreds of keV/u rather than a few keV/u. Olson's suggestion was based on a theoretical study of intermediate energy $H^+ + H$ collisions. His calculations, based on classical methods, indicated an unexpectedly large number of target electrons being ejected with velocities near $v/2$ and predicted that this feature occurs primarily, but not exclusively, in the forward direction. Additional theoretical work based on quantum mechanical methods [*Wang* and *Burgdörfer* (1989), *Wang* et al. (1991)] yielded controversial results so that the discussion of saddle point electron emission remained with open questions.

Irby et al. (1993) suggested that an expected signature of this mechanism is a maximum in the differential cross section, $d^2\sigma/dv_e d\Omega$, at an ejected electron velocity, v_e, given by

$$v_e = \frac{v}{1 + (q/q_t)}, \qquad (5.8)$$

where q and q_t are, respectively, the projectile and target ion charge states after the collision. As usual, v is the projectile velocity. Thus, experimental efforts to confirm the presence of saddle point ionization have centered on looking for evidence of this maximum and investigating its position as a function of projectile charge state. These studies have been performed by groups associated with the University of Missouri-Rolla, Centro Atómico Bariloche, and the Pacific Northwest Laboratory.

Most of these studies have concentrated on ionization induced by H^+ and He^{2+} ions having energies between 50 and 100 keV/u. An early experimental study of saddle emission was performed by *Meckbach* et al. (1986). Although it was later found that the experimental data required revision, this study pioneered the experimental efforts to confirm the existence and importance of saddle point ionization. Later studies by *Olson* et al. (1987), *Irby* et al. (1988) and *Gay* et al. (1990) were interpreted as supporting the hypothesis that saddle point ionization is an important mechanism. Moreover, *Irby* et al. (1993) used partially stripped carbon ions to extend the investigation to higher charge state projectiles. These data also supported the interpretation that saddle point ionization is important at intermediate impact velocities.

However, later studies by the Bariloche group lead to antipodal interpretations. The experimental results by *Bernardi* et al. (1989, 1990) and *Meckbach* et al. (1991) did not confirm the earlier conclusions by Meckbach et al. (1986). Moreover, studies by *DuBois* (1993, 1994) have yielded data that questions the studies supporting the hypothesis that saddle point ionization is an important mechanism in the 50–100 keV/u impact energy range.

The controversy centers on two points, namely i) whether the method used to display the data artificially introduces a signature of saddle point ionization, and ii) whether the experimental data which support the saddle point ionization hypothesis are reliable. Concerning the first point, the supporters of saddle point ionization maintain that the differential electron emission should be studied in ejected electron velocity space, $d^2\sigma/dv_e\,d\Omega$. Displaying the data in this fashion will inevitably introduce a maximum because $d^2\sigma/dv_e\,d\Omega$ is related to the doubly-differential cross section, $d^2\sigma/d\epsilon\,d\Omega$, by

$$\frac{d^2\sigma}{dv_e\,d\Omega} = v_e \frac{d^2\sigma}{d\epsilon\,d\Omega}. \tag{5.9}$$

It is obvious that as $v_e \to 0$, $d^2\sigma/dv_e\,d\Omega$ tends to zero. At the other extreme, large v_e, $d^2\sigma/dv_e\,d\Omega$ tends to zero, too. As a result, $d^2\sigma/dv_e\,d\Omega$ will always maximize at some intermediate value of v_e.

Meckbach et al. (1991) suggested that it is more appropriate to plot the cross section in v_e space using $d^2\sigma/dv_e$ where $dv_e = v_e^2\,dv_e\,d\Omega$ is the velocity volume element. They demonstrated that when the data are displayed in this fashion, no enhancement in the region near $v/2$ is indicated. They furthermore concluded that what has been interpreted as "unambiguous evidence" of saddle point emission [*Gay* et al. (1988)] is merely the remnants of the intense electron capture to the continuum cusp at non-zero degree forward angles. Moreover, *Bernardi* et al. (1990) addressed the issue of the saddle point maximum shifting toward smaller v_e with increasing projectile charge. Such a shift is predicted and had been reported. However, their measurements failed to confirm such a shift.

To help resolve this controversy, *DuBois* (1993) performed an additional measurement of the doubly-differential electron emission for H^+ and He^{2+} impact on a helium target. From his and the other data available for these

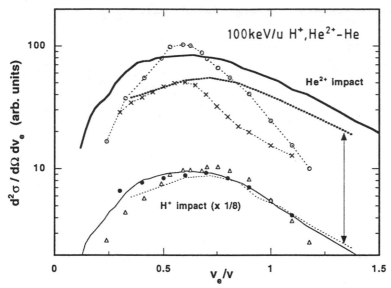

Fig. 5.26. Double-differential electron emission for $100\,\text{keV/u}$ H^+ and He^{2+} impact on helium. The H^+ / He^{2+} experimental data are for $15°$ emission, *DuBois* (1993) (*thin* and *thick solid curves*); $17°$ emission, *Bernardi* et al. (1989) (*thin* and *thick dashed curves*); $15°$ emission obtained by interpolation, *Gay* et al. (1990) ($\bullet$, $\circ$); and $17°$ emission, *Irby* et al. (1988) ($\triangle$, $\times$). For comparison, differences in absolute normalization have been removed by multiplying the *Bernardi* et al. (1989), *Gay* et al. (1990), and *Irby* et al. (1988) data by 0.65, 0.69 and 0.44 respectively. The ordinate is the electron velocity divided by the projectile velocity. The *arrow* at the right side of the figure indicates a relative difference which would be expected from Z_p^2 scaling taking into account that the proton impact data have been shifted downward by a factor of 8 for display purposes

systems, he concluded that the data demonstrating a shift in the cross section maximum are in error. As illustrated in Fig. 5.26, four independent experimental investigations of the differential electron emission in $H^+ + He$ collisions are in agreement, except for differences in absolute magnitudes, and all four investigations indicate a maximum in the region where expected. However, differences in the He^{2+} impact data from these same research groups are obvious. Most importantly, two of the studies observe that the maximum in $d^2\sigma/dv_e\,d\Omega$ shifts toward smaller v_e with increasing projectile charge, whereas two do not.

Based on a comparison of ratios of the differential cross sections for He^{2+} and H^+ that were reported by the various groups, *DuBois* (1993) suggested that the two data sets demonstrating a shift contain flawed He^{2+} impact data. At high impact energies, first order perturbation theory predicts that the cross section ratio should be 4, as confirmed by DuBois (1993). However, as shown in Fig. 5.27 for $100\,\text{keV/u}$ impact, the data of *Bernardi* et al. (1990) yield a ratio near 4 for fast electron emission and a smaller ratio for

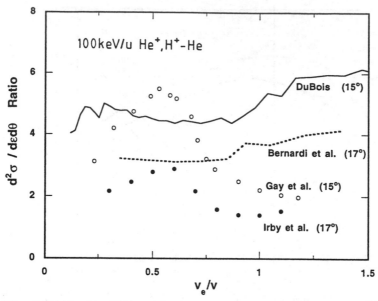

Fig. 5.27. Double-differential cross section ratios for $100\,\mathrm{keV/u}$ He^{2+} and H^+ impact on helium. The data and emission angles are those shown in Fig. 5.26

slow electron emission. The ratios of *DuBois* (1993) agree in shape but are systematically larger. However, the ratios of *Irby* et al. (1988) and *Gay* et al. (1990) are considerably different in shape and are considerably smaller than 4 for close collisions, i.e., as $v_e/v \to 2$. Since all four sets of proton data nearly agree, this strongly implies that the collision systems which demonstrate a shifted maximum are others than the system $\mathrm{He}^{2+} + \mathrm{He}$.

In a following study, *DuBois* (1994) performed a more extensive investigation using partially stripped carbon ions. These data again do not confirm specific features of the data of *Irby* et al. (1993) which appeared to support the saddle point ionization picture. More importantly, *DuBois* (1994) suggested that partially stripped ions are inappropriate for investigating the saddle point ionization phenomenon. This is because partial screening of the projectile nuclear charge by its bound electrons will inherently lead to features strongly similar to those expected for saddle point ionization, i.e., screening of the nuclear charge by the projectile electrons suppresses the emission of low-energy electrons whereas the emission of energetic electrons is relatively unaffected. The suppression at low energies due to screening increases as bound electrons are added to a bare projectile ion (Sect. 6.2). The result is that a maximum in $d^2\sigma/dv_e\,d\Omega$ is produced and this maximum shifts toward smaller v_e with increasing net projectile charge, precisely as expected for saddle point ionization. Therefore, *DuBois* (1994) suggested that the search for saddle-point ionization should be restricted to bare ions, or at least to highly stripped ions, where screening effects can be ignored.

To summarize, experimental studies performed in the 50–150 keV/u region in order to investigate whether saddle point ionization is an important mechanism at higher impact energies have yielded conflicting data and interpretations. From the information available, we conclude that there is no definitive experimental evidence that saddle point ionization is a significant mechanism in energetic collisions. However, we remind the reader that at extremely low impact energies saddle-point ionization has been shown to be important. As low impact energies are out of the scope of this review, these studies will not be discussed. For details, the reader is referred to the recent studies of *Macek* and *Ovchinnikov* (1994) and *Pieksma* et al. (1994).

5.4 Saturation Effects

The Coulomb force exerted on a target electron by a bare projectile ion is proportional to the perturbation strength Z_p/v where Z_p is the projectile charge and v is its velocity. When the interaction is weak, perturbation theories are appropriate for calculating the ionization cross sections. For example, in the Born approximation a series expansion of the ionization amplitudes can be performed where the terms are proportional to $(Z_p/v)^n$, n being the order of the expansion. For weak interactions, e.g., for large v and small Z_p, only the first-order term is important and the total and differential cross sections scale quadratically with the force, i.e., as $(Z_p/v)^2$. Hence, heavy ion impact cross sections are typically obtained in the Born approximation by multiplying calculated proton impact cross sections by $(Z_p/v)^2$, a feature that has repeatedly been applied throughout this review. Using similar logic, in the semiclassical approximation [*Hansteen* (1975)] the proton impact ionization probabilities, $P(b)$, are scaled by $(Z_p/v)^2$ in order to determine cross sections for heavy ion impact.

However, the underlying condition for Z_p^2 scaling of ionization cross sections is that the interaction is weak. This means that, for a given v, Z_p cannot be increased indefinitely without higher-order terms in the Born expansion becoming important. Ultimately, for large enough Z_p, the first-order theory violates unitarity conditions as described, e.g., by *Landau* and *Lifshitz* (1958). As a consequence, for strong interactions the cross sections increase at a rate slower than predicted by Z_p^2 scaling. We shall refer to this weaker scaling dependence as saturation. (Note that this definition excludes deviations from $(Z_p/v)^2$ scaling where the cross sections become larger.)

Several studies [*Bohr* and *Lindhard* (1954), *Schlachter* et al. (1981), *Knudsen* et al. (1984)] have demonstrated the weaker dependence on Z_p when the interactions become strong. These studies showed that the ionization cross sections for many different collision systems fall on a universal curve if both the cross sections and the impact velocities are scaled linearly by Z_p, i.e., σ/Z_p plotted versus v^2/Z_p yields a universal curve.

It is important to remember several aspects about this scaling picture. First, due to Z_p/v scaling, the division between the weak- and the strong-interaction regions depends on the impact velocity, e.g., as the impact velocity increases, the weak-interaction region extends to higher projectile charge states. Second, because total cross sections have demonstrated saturation effects, it follows that deviations from Z_p^2 scaling should also be observed in the differential electron emission cross sections. Third, these generalized statements apply only to situations where single, outer shell ionization dominates and the ionizing projectile is a fully, or highly, stripped ion where screening effects are minimal.

Saturation effects may be associated with a single center that is strongly affected by an external interaction. As will be shown, two-center effects leading to a redistribution of the electron emission may be accompanied by saturation effects which lead to an emission strength that is reduced from that expected from Z_p^2 scaling. But, depending upon various factors, a projectile charge may be large enough to produce an observable redistribution of the electron emission without an observable suppression in the overall emission.

5.4.1 Differential Cross Sections

It was suggested above that saturation effects should be observed in the differential electron emission cross sections. As was shown in the preceding section (Figs. 5.22–24,for fast, highly charged bare ion impact, the low-energy electron emission cross sections are suppressed with respect to scaled proton cross sections when the emission angle is 90° or larger. This feature was discussed in terms of two-center effects where the electron emission is redistributed from backward to forward emission angles. However, as shown in Fig. 5.28, for fast, highly charged ion impact the electron emission is not only redistributed but the total emission is also suppressed from values expected for Z_p^2 scaling.

For this demonstration, we evaluated singly-differential cross sections, SDCS, which were obtained by multiplying the scaled CDW-EIS SDCS ratios of *Fainstein* et al. (1991) with the absolute proton impact cross sections of *Rudd* et al. (1976), and integrating over the electron energy. The results were then multiplied by $\sin\theta$. Thus, the area under the curves is the total ionization cross section divided by Z_p^2. The figure clearly illustrates that with increasing projectile charge, the scaled emission is slightly enhanced for forward angles but is significantly reduced for intermediate and large angles, with the overall effect being a reduction in the scaled total ionization cross section.

It is recalled that for decreasing impact energies, deviations from $(Z_p/v)^2$ scaling will also occur for smaller values of Z_p. This is demonstrated in Figs. 5.29,30 using 0.1 MeV/u helium ion impact data of *Bernardi* et al. (1989). In these figures, differential cross sections for He^{2+} impact are compared with proton impact cross sections which have been scaled by Z_p^2. Note

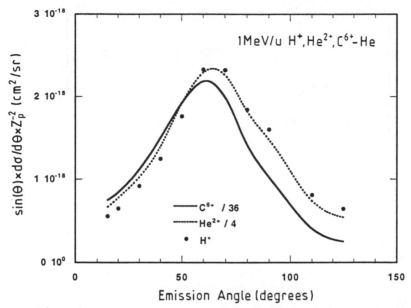

Fig. 5.28. Singly-differential cross sections for 1 MeV/u H$^+$, He^{2+}, and C^{6+} impact on helium. The cross sections, from the work of *Bernardi* et al. (1989) and *Pedersen* et al. (1990), have been multiplied by $\sin(\theta)$ and divided by the square of the projectile nuclear charge in order to demonstrate how two-center and saturation effects influence the total electron emission

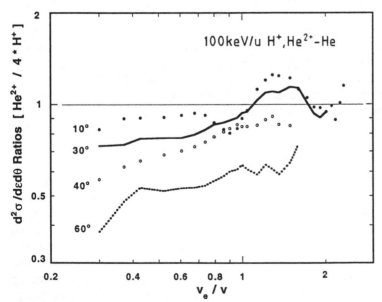

Fig. 5.29. Ratios of doubly differential cross sections for 0.1 MeV/u H$^+$ and He^{2+} impact on helium. The cross sections of *Bernardi et al.* (1989) are displayed as in Fig. 5.28

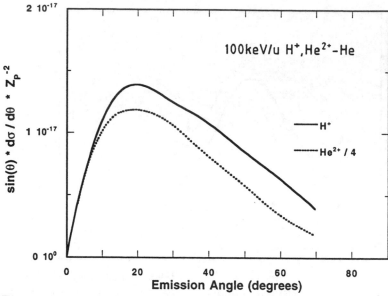

Fig. 5.30. Singly differential cross sections for 0.1 MeV/u H⁺ and He²⁺ impact on helium. The cross sections of *Bernardi* et al. (1989) are displayed as in Fig. 5.28

that at this lower impact energy the scaled electron emission is suppressed for all angles when the emission velocity is less than the impact velocity, as was discussed in the previous section (Fig. 5.26).

Also, note that in contrast to what was observed for higher impact energies, the enhancement of the scaled emission of energetic electrons is relatively small. Thus, as shown in Fig. 5.30, at lower impact energies no enhancement in the forward differential electron emission is observed. This implies that two-center effects (which lead to an enhancement in the forward direction) are likely to be masked by saturation effects.

5.4.2 Total Cross Sections

As demonstrated, saturation effects influence the overall electron emission. Therefore, many studies of the scaling of total ionization cross sections have been performed and quadratic or linear scaling dependencies as a function of projectile charge are tested. In the mid 1980s, *Knudsen* et al. (1984) investigated single and double ionization of helium by various highly stripped ions and demonstrated that as the impact energy decreased, deviations from Z_p^2 scaling occur at ever decreasing values of Z_p.

This is illustrated in Fig. 5.31 where theoretical cross sections for ionization of helium are compared for various projectiles. The data were obtained using the CDW-EIS code by *Fainstein* et al. (1991). In Fig. 5.31a, the cross

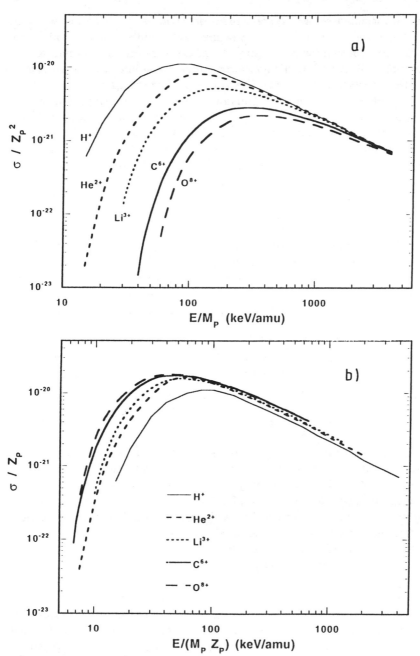

Fig. 5.31a,b. Total cross sections for ionization of helium by fully stripped ion impact calculated using the CDW-EIS of *Fainstein* et al. (1991). In (a) the cross sections are divided by the square of the projectile nuclear charge. In (b) both cross sections and impact energy are divided by the projectile nuclear charge. Note the improvement of the scaling procedure

sections are divided by Z_p^2. Where weak-interactions dominate, a universal curve is expected. It is seen that Z_p^2 scaling is invalid below approximately 0.4, 1, 2, and 4 MeV/u for fully stripped helium, lithium, carbon and oxygen impact, respectively. Below these impact energies, the interactions are strong so that they cannot be treated in first order perturbation theory.

As mentioned earlier, in the regime of stronger perturbations, a universal curve is expected if both the cross sections and the impact velocities are divided by Z_p [Bohr and Lindhard (1954)]. For this type of scaling a universal curve is also expected at high-impact energies because the cross sections have an approximate $Z_p^2 E_p^{-1} \ln E_p$ dependence (2.2). The theoretical cross sections from Fig. 5.31a were scaled in this fashion and are shown in Fig. 5.31b. It is seen that the agreement for the heavier ion impact data is extended to lower impact energies. Nevertheless, a near perfect coincidence of the data is achieved only for the C^{6+} and O^{8+} projectiles. At small values of $E_p/M_p Z_p$, the deviations in the scaled data for the various collision systems have been attributed to binding effects [Fainstein et al. (1987)].

Finally, we discuss attempts to factorize cross sections including saturation effects. Gillespie (1983) proposed a scaling which represents saturation by a factorized function depending on the parameter Z_p/v^2, i.e.,

$$\sigma(Z_p, v) = Z_p^2 f(Z_p/v^2) \sigma(v), \qquad (5.10)$$

where $\sigma(v)$ is the cross section obtained by perturbation theory for small Z_p and $\sigma(Z_p, v)$ is the corresponding cross section attributed to an arbitrary Z_p. The function containing the effect of saturation is given as

$$f(Z_p/v^2) = exp\left(-\lambda \frac{Z_p}{v^2}\right), \qquad (5.11)$$

where λ is an adjustable parameter. Gillespie (1983) found good agreement for total ionization cross sections by fully stripped ions incident on H, H_2, and He using $Z_p = 1 - 13$ and projectile energies from 0.075 to 5 MeV/u. The parameter λ was determined to be 0.76 for H, 1.0 for H_2, and 2.0 for He. Similar work has been performed by Shah and Gilbody (1983) showing that the factorization formula (5.10) also holds for dressed projectiles.

More recently, Colavecchia et al. (1995) has applied the factorization formula (5.10) to singly differential cross sections. Instead of using theoretical cross sections $\sigma(v)$ from perturbation theory, they based their analysis on semiempirical cross sections $\sigma(Z_p = 1, v)$ given by Rudd (1988) for proton impact. The scaling results show good agreement with various experimental data from Platten et al. (1987) and Pedersen et al. (1991). Hence, empirically, the factorization of cross sections including saturation has been proven to be a useful procedure. It allow for the conclusion that saturation phenomena are rather independent on the details of the electron production mechanisms.

Summarizing, with increasing Z_p/v the interaction with the target electron becomes strong and the cross sections will saturate, meaning that the

Z_p^2 scaling breaks down. This saturation primarily reduces the lower-energy electron emission, relative to that expected from scaling proton impact cross sections by Z_p^2. Saturation effects will occur at any impact velocity. As was shown, at $100 \, \mathrm{keV/u}$, interaction strengths are sufficiently strong that Z_p^2 scaling never applies.

6. Ionization by Dressed Projectiles

In ionization by dressed particles, the bound electrons play essentially two different roles. For a pictorial description of these roles, refer back to Fig. 2.6 in Sect. 2.5. One role of the bound electrons is primarily passive, meaning that they alter the Coulomb fields of the collision partners by partially screening their respective nuclear charges. In this case, ionization of one of the collision partners results from an interaction between its electron and the screened nucleus of the other partner. The other role that bound electrons play is active, meaning that electrons bound to one collision partner interact with those bound to the other partner. The massive nuclei participate in the interaction only in the sense that they provide a distribution of momenta for the bound electrons.

These passive and active roles influence the ionization cross sections in opposite ways. The passive (screening) role reduces the cross section with respect to that for a bare nuclear charge. In contrast, the active role increases the cross section with respect to that for a bare nucleus. It should be kept in mind that both roles are present for dressed particle impact. However, either may dominate and their relative strengths depend on the impact parameters, on the relative magnitudes of the projectile and target nuclear charges, and on the number of electrons bound to each center.

6.1 Definitions and Notations

In the work of *Bates* and *Griffing* (1955), the passive and active roles were referred to as "single-" and "double-transitions". However, renewed interest at various times has resulted in many different names being attached to these processes. The passive role has been referred to as an "electron-nuclear" $(e-n)$ interaction, as a "monoelectronic" process, or as the "singly inelastic" channel. The active role has been referred to as an "electron-electron" $(e-e)$ interaction, as a "two-center electron correlation", as a "two-center dielectronic" interaction, as the "doubly inelastic" channel or as "antiscreening". For consistency, we shall use the term "screened nuclear" or "monoelectronic" interaction, designated by "$e-n$", when referring to the passive role and "two-center dielectronic" interaction, designated by "$e-e$", when referring to the active role. When misunderstanding can be excluded, the term two-center dielectronic is simply referred to as dielectronic.

Thus, the presence of electrons bound to the incoming projectiles leads to many additional interaction channels. This is illustrated below using energetic hydrogen atom impact on helium. For illustrative purposes, we have ignored double electron transitions on the helium target.

$$
H + He \rightarrow
\begin{cases}
\begin{rcases}
H + He* & (1) \\
H + He^+ + e_T & (2)
\end{rcases} e_T - n_P \\[2ex]
\begin{rcases}
H* + He & (3) \\
H^+ + He + e_P & (4)
\end{rcases} e_P - n_T \\[2ex]
\begin{rcases}
H* + He* & (5) \\
H* + He^+ + e_T & (6) \\
H^+ + He* + e_P & (7) \\
H^+ + He^+ + e_T + e_P & (8)
\end{rcases} e_P - e_T \ .
\end{cases}
\tag{6.1}
$$

Note that in the various interactions we have explicitly designated components of the target and projectile by T and P, respectively. Also it is important to keep in mind that independent, higher-order $e - n$ interactions, e.g., $(e_T - n_P)(n_T - e_P)$, can also lead to mutual excitation and ionization of the collision partners. However, the dielectronic processes of interest are first-order interactions and, hence, should prevail at high impact energies.

6.2 Screening Effects of Projectile Electrons

The most important effect of a passive electron carried by the projectile is the screening of the incident nuclear charge. A passive electron produces a mean field added to the Coulomb field of the projectile. Hence, as a result of the combined effects of the projectile charge and the bound electrons, a screened Coulomb field, is created. This field depends on the states of the projectile electrons.

In the following, the attempt is made to provide the basic understanding of screening phenomena in electron emission processes. It will be shown that the energy of the ejected electron is a most sensitive parameter for verifying screening effects. It governs the minimum momentum transfer in the collision which, in turn, is the most important parameter that determines screening effects. Hence, it will be shown that differential cross sections provide an essential tool to obtain information about screening mechanisms.

6.2.1 Screened Point Charge

In ion-atom collisions, it is common practice to treat screening effects by the introduction of an effective projectile charge $Z_p^{eff} \leq Z_p$ that behaves like a

Coulombic point charge. Consequently, the double-differential cross section that scales with the square of the projectile charge is obtained as

$$\frac{d\sigma}{d\Omega\, d\epsilon} = \left(Z_p^{eff}\right)^2 \frac{d\sigma(1)}{d\Omega\, d\epsilon}, \tag{6.2}$$

where $d\sigma(1)/d\Omega\, d\epsilon$ is the corresponding cross section for proton impact. Equation (6.2) should be taken as a definition of the effective charge Z_p^{eff}.

The effective charge can also be estimated on the basis of simple models. Since the early beginning of atomic physics, Slater's screening rules have been adopted as a useful tool. Furthermore, a dressed projectile field has often been described by a Bohr potential involving an exponential screening. Electron scattering by an exponentially screened potential is described by the Rutherford formula (3.3) given in Sect. 3.1.1. This formula clearly indicates that the bound electrons have the effect of reducing the nuclear charge. However, care should be taken with the approximate treatments of screening as they may introduce considerable uncertainties into the evaluation of the cross sections. Nevertheless, approximate procedures are often applied since the adequate treatment of the screening effects is still a difficult problem in ion-atom collisions.

The major work concerning screening effects in ionization has been conducted with respect to total cross sections, in particular, to those associated with the removal of inner shell electrons [*Meyerhof* and *Taulbjerg* (1977)]. The effective charge depends on various experimental parameters such as the projectile velocity and the binding energy of the ejected electron. In contrast to the extensive studies of total cross sections, the amount of work devoted to double-differential cross sections for electron emission is limited. This is partially due to the fact that the data analysis is more complex. For double-differential cross sections the number of experimental parameters is increased. To analyze screening effects on electron emission, the variation with the energy and angle of the ejected electron has to be considered. In this case the concept of an effective projectile charge loses significance as it becomes dependent on parameters associated with the target.

6.2.2 Experimental Studies

In accordance with (6.2), it is useful to compare measured ionization cross sections for dressed projectiles with the corresponding cross sections for a bare projectile. Early studies of cross section ratios for electron emission by dressed and bare projectiles have been performed by *Wilson* and *Toburen* (1973) using H_2^+ and H^+ impact at energies of about 1 MeV/u. Similar work with light projectiles has been carried out by *McKnight* and *Rains* (1976) and *Sataka* et al. (1979a,b). These studies revealed the pronounced electron-loss peak that is due to the removal of projectile electrons. Unfortunately, the analysis of the screening effects is obscured by the appearance of the electron-loss peak so that finer details are lost. *Wilson* and *Toburen* (1973)

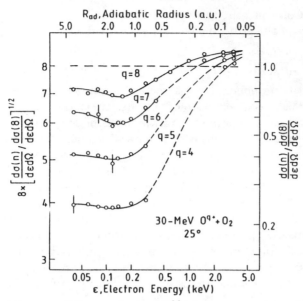

Fig. 6.1. Effective charge Z_p^{eff} of the projectile as defined by the square root of ratios of DDCS for partially stripped oxygen projectiles, with respect to the corresponding cross sections for O^{8+} in collisions of 30-MeV O^{q+} on O_2. The emission angle is $25°$. The data are based on cross section from *Stolterfoht* et al. (1974,1994)

noted significant variations of the cross section ratio in the soft collision region, however, no attempt was made to analyze them in terms of projectile screening.

Screening effects became more obvious in a series of measurements dealing with heavy projectiles whose charge state was varied over a wide range [*Stolterfoht* et al. (1974), *Stolterfoht* (1978)]. A few results of these experiments have already been discussed in Sect. 2.5. Looking back at Fig. 2.5 it is noted that different parts of the energy spectra exhibit different dependencies on the charge state of the O^{q+} projectile incident on O_2. For instance, the intensity of the binary-encounter peak barely changes with the incident charge state whereas the soft-collision electron exhibits significant variations.

This behavior can be studied in more detail when the corresponding cross section ratios are considered. Figure 6.1 shows electron emission cross sections for different charge states q of the projectile in relation to the data for bare O^{8+} ions. To avoid the electron-loss peak obscuring the analysis, the cross section ratios have been interpolated through the range where the electron-loss peak appears. The right ordinate refers to cross section ratios, whereas, the left ordinate refers to the effective charge

$$Z_p^{eff} = 8 \left(\frac{d\sigma(q)}{d\epsilon\, d\Omega} \Big/ \frac{d\sigma(8)}{d\epsilon\, d\Omega} \right)^{1/2}$$

(6.3)

which is based on (6.2). The cross section $d\sigma(q)/d\epsilon\,d\Omega$ refers to the O^{q+} projectile. Figure 6.1 shows that Z_p^{eff} varies significantly as the energy of the electrons changes.

The decisive parameter to interpret the data is the adiabatic radius $R_{ad} = 1/K_m$ which is equal to the inverse minimum momentum transferred during the collision. The decisive influence of $R_{ad} = 1/K_m$ on the screening follows from the Born approximation (see further below) and has been verified experimentally [Stolterfoht (1978)]. It should be kept in mind that the adiabatic radius is conceptionally different from the internuclear distance. The latter quantity measures the distance between the two heavy particles, whereas the adiabatic radius is a measure for the distance between the incident ion and the target electron. The two quantities are similar for glancing collisions, however, they differ significantly when the collision partners approach each other. Nevertheless, to obtain small values for the adiabatic radius, the projectile must penetrate deep within the electron cloud of the target atom where the screening effects are small. Hence, a functional dependence is expected between the adiabatic radius and the screening effects. For small radii, screening effects are negligible whereas they gain importance for large adiabatic radii.

The relation between the energy ϵ of the ejected electron and the adiabatic radius, or the inverted minimum momentum transfer, is given by (Sect. 3.4.1)

$$R_{ad} = \frac{1}{K_m} = \frac{v}{\epsilon + E_b}. \tag{6.4}$$

As before, v is the projectile velocity, ϵ is the electron energy, and E_b is the ionization energy of the target electron. It is seen that for increasing electron energy the adiabatic radius decreases without limitation. On the contrary, for decreasing electron energy the adiabatic radius approaches a finite value governed by the minimum energy transfer E_b (i.e., the ionization energy). The maximum value of the adiabatic radius corresponds to the minimum momentum transfer, i.e., a quantity well known within the framework of the Born approximation.

The adiabatic radius R_{ad} is shown along the x-axis at the top of Fig. 6.1. It varies within two orders of magnitude in accordance with the variation of the electron energy. Thus, extreme values for screening can be studied at the limits of the energy range used here. We note that it would be difficult to cover such a two-orders-of-magnitude range in experiments where the impact parameter is varied. Hence, unique information about screening effects is achieved from the present measurements of energy differential cross sections for electron emission.

Figure 6.1 shows that the projectile nucleus is fully screened for adiabatic radii larger than $\sim$1 a.u. In this region of the soft-collision electrons the effective nuclear charge Z_p^{eff} is about equal to the charge state q of the projectile. As R_{ad} decreases beyond 1 a.u., the screening becomes less important and the minimum value is reached at about 0.1 a.u. In this region of the binary-encounter peak, the projectile has entered into the K orbital of the target atom oxygen. Thus, screening is expected to be negligible. Indeed it is found that the effective charge Z_p^{eff} is about equal to the nuclear charge $Z_p = 8$ of the projectile. This shows that the screening effects reach their extreme values when the electron energy is varied from the soft-collision to the binary-encounter region.

In the early 1980s several studies of double-differential cross sections for electron emission were devoted to screening effects. The Debrecen group [*Köver* et al. (1980, 1982)] performed detailed experiments of cross section ratios for He^+/H^+ and H_2^+/H^+ impact on Ar. They found significant screening effects which they interpreted in terms of the variation of the adiabatic radius which was assumed to be equal to a mean impact parameter. Furthermore, screening effects were observed in reversed collision systems where the ionization of the projectile is influenced by the passive electrons of the neutral target atom [*Prost* et al. (1982), *Schneider* et al. (1983), *Köver* et al. (1988)].

In recent years the interest in screening effects has diminished since the attention has turned to higher-order effects of the mean field formed by the passive electrons. Figure 6.1 shows that in the binary encounter region (2–4 keV), the cross sections for dressed projectiles are enhanced with respect to those for the bare ion. This enhancement is rather weak and was not noted prior to revising the early data [*Stolterfoht* et al. (1994), *Zouros* et al. (1994)]. The present data are due to an electron observation angle of 25°. Near 0° the enhancement of the binary-encounter peak becomes significant [*Richard* et al. (1990)]. This observation showed that due to higher-order effects, the passive electrons do not always decrease the cross sections for electron emission. Higher-order phenomena of the mean projectile field will be discussed in detail in a later section. Here, we shall first discuss the theoretical aspect of electronic screening.

6.2.3 Theoretical Interpretation

As noted before the theoretical treatment of screening dates back to the early work by *Bates* and *Griffing* (1953), who studied the interactions of electrons bound at the projectile and target atom. As shown in Sect. 3.7.1, their analysis is based on the first Born approximation yielding expression (3.211) for the ionization cross section [*Briggs* and *Taulbjerg* (1978), *McGuire* et al. (1981)]. This expression for the double-differential cross section is obtained as

$$\frac{d\sigma_{sc}}{d\epsilon\,d\Omega} = k\frac{8\pi}{v^2}\int_{K_m}^{K_M} |Z_p - S_p(K)|^2\,|F_{o,k}(-K)|^2\,\frac{dK}{K^3}, \tag{6.5}$$

where $F_{o,k}(K)$ is the inelastic form factor for electron emission from the target and K_m is the minimum momentum transfer. The remaining notation is specified in Sect. 3.7.1.

The screening function $S_p(K)$ describes the electrons bound at the projectile. Without this term, expression (6.5) reduces to the well-known result of the Born approximation for bare projectiles (Sect. 3.4.2). For projectiles carrying one electron only, the screening function $S_p(K)$ is equal to its elastic form factor $F_{i',i'}(K)$. For multielectron projectiles, the contributions from different shells are added coherently

$$S_p(K) = \sum_s N_p^{(s)} F_{i',i'}^{(s)}(K), \qquad (6.6)$$

where $N_p^{(s)}$ is the number of electrons in the shell labeled s.

The contributions from the projectile nucleus and the electrons may be combined to a net projectile charge Q_p that depends on the momentum transfer K

$$Q_p(K) = Z_p - S_p(K). \qquad (6.7)$$

Hence, the effective charge Z_p^{eff} defined by (6.2) is obtained as a mean value of $Q_p(K)$ weighted with the inelastic form factor of the target

$$Z_p^{eff} = \left[\frac{\int\limits_{K_m}^{K_M} |Q_p(K)|^2 \, |F_{o,k}(-K)|^2 \, \dfrac{dK}{K^3}}{\int\limits_{K_m}^{K_M} |F_{o,k}(-K)|^2 \, \dfrac{dK}{K^3}} \right]^{\frac{1}{2}}. \qquad (6.8)$$

Intuitively, the effective charge Z_p^{eff} is associated with properties of the projectile. Regarding the denominator in (6.8) we note that the terms describing the projectile and target properties factorize in the integrand. Thus, indeed, $Q_p(K)$ is uniquely determined by projectile properties. However, when the mean value Z_p^{eff} is evaluated, this factorization is lost due to the integration over the momentum transfer K. Consequently, Z_p^{eff} is affected by target properties, i.e., in particular it may depend on the energy and angle of the ejected electron.

This dependence has been studied theoretically by *Manson* and *Toburen* (1981) who determined double-differential cross sections for ionization of He by dressed and bare projectiles. In Fig. 6.2 the results for 0.5-MeV/u H^+, He^+, and He^{2+} impact are compared. In the calculations, Hartree-Fock-Slater wave functions were used for the target. The electron of the He^+ projectile is described by a hydrogen-like 1s wave function. It is seen that at low electron energies the He^+ cross section approaches that of H^+ whereas it tends to the data of He^{2+} at high energies. In these regions, the extreme cases of complete and negligible screening are achieved. In the intermediate range a smooth transition between the two asymptotic cases is observed.

To analyze this behavior quantitatively, we consider the explicit expression of the elastic form factor $F_{i',i'}(K)$ which is obtained as the Fourier trans-

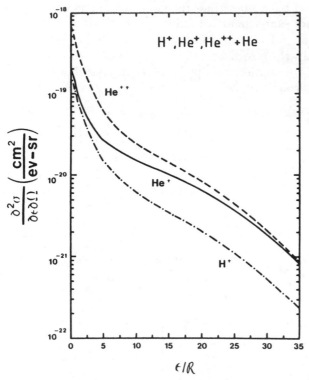

Fig. 6.2. Double-differential cross sections for ionization of He by equal velocity H^+, He^+, and He^{2+} as a function of the ejected electron energy (in units of Rydbergs = 0.5 a.u.) at an ejection angle of 60°. The He^+ electron is assumed to be passive; it remains in the ground state. The incident energy is 0.5-MeV/u. The data are calculated by *Manson* and *Toburen* (1981) using the Born approximation

form of the mean field produced by the projectile electrons (Sect. 3.7.1). For a hydrogenic 1s orbital it follows that

$$F_{1s',1s'}(K) = \frac{1}{\left[1 + (K/2\alpha_p)^2\right]^2}, \tag{6.9}$$

where $\alpha_p = \sqrt{2E_b}$ is associated with the mean velocity of the bound projectile electron. It is noted that $F_{1s',1s'}(K) \to 1$ as $K \to 0$, i.e., the screening effects maximize for small K. On the other hand, for $K \to \infty$ it follows that $F_{1s',1s'}(K) \to 0$, i.e. the screening effects disappear for large K. In Fig. 6.2 these cases correspond to the soft- and binary-collision regions, respectively.

Equation (6.9) can be used to evaluate an approximate expression for Z_p^{eff}. It is understood from (6.8) that the integrands depend on K^{-3} so that they are strongly enhanced at the minimum value $K \approx K_m$. Therefore, a peaking approximation can be applied replacing the variable $Q_p(K)$ by the constant $Q_p(cK_m)$ which involves the (effective) momentum transfer cK_m

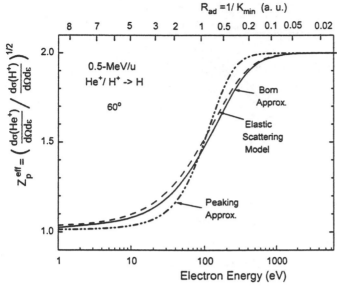

Fig. 6.3. Ratio of cross sections for ionization of He by 0.5-MeV/u He$^+$ to H$^+$ impact as a function of the electron energy. On the *top* of the figure corresponding results for the adiabatic radius are given. The Born approximation data are based on the on the results shown in Fig. 6.2 [*Manson* and *Toburen* (1981)]. The peaking approximation data are obtained from 6.10. The elastic scattering results are from 6.11

where $c > 1$ is an adjustable parameter. From (6.8) it follows that $Z_p^{eff} = Q_p(cK_m)$ and, hence,

$$Z_p^{eff} = Z_p - \frac{1}{\left[1 + (cK_m/2\alpha_p)^2\right]^2}. \tag{6.10}$$

This expression confirms the previous assumption that the effective nuclear charge Z_p^{eff} is a function of the minimum momentum transfer K_m and, thus, of the adiabatic radius R_{ad}, *Stolterfoht* (1978). In particular, it is important to note that Z_p^{eff} is not explicitly dependent on the projectile velocity. The velocity dependence is included in the momentum transfer K_m.

To study effective charges in detail, cross section ratios for the projectiles He$^+$ and H$^+$ are plotted in Fig. 6.3. The results clearly depict the transition from the region of full screening, $Z_p^{eff} = 1$, to the region of negligible screening, $Z_p^{eff} = 2$, for the He$^+$ projectile. The data obtained from the calculations by *Manson* and *Toburen* (1981) in (Fig. 6.2) are compared with model results from the approximate formula (6.10) using $c = 2.8$. The two sets of data agree to within $\pm 10\,\%$ which is rather satisfactory in view of the simplicity of the approximate formula.

When inducing electron emission, screening effects may not only depend on the energy but also on the angle of the ejected electron [*Manson* and

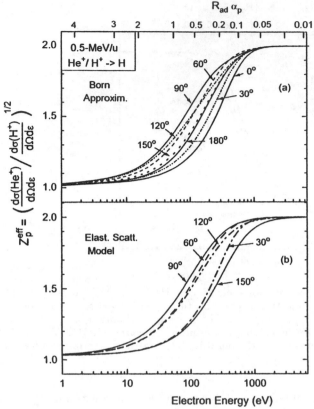

Fig. 6.4a,b. Ratio of cross sections for ionization of H by 0.5-MeV/u He$^+$ impact as a function of the electron energy. On the *top* of the figure, corresponding results for the adiabatic radius are *given*. The observation angle is indicated at the curves. In (**a**) the results are calculated using 6.8 in conjunction with the analytic form factor formula given by *Landau* and *Lifschitz* (1958). In (**b**) the data are calculated using 6.11

Toburen (1981)]. To verify this possible dependence, we calculated Z_p^{eff} for 0.5-MeV/u He$^+$ and H$^+$ impact on H using (6.8). The form factors were evaluated by means of an analytic formula (3.99) from *Landau* and *Lifschitz* (1958). Results for various electron observation angles are shown in Fig. 6.4a. It is seen that for 0° the variation of the Z_p^{eff} curve occurs at relatively high electron energies. The Z_p^{eff} curve shifts to lower energy with increasing angle and it reaches a minimum for 90°. Then, the Z_p^{eff} curve shifts back to higher energy as the angle increases until 180°.

The variation of screening with the electron emission angle is significant. Recalling that the cross sections depend on the square of Z_p^{eff} it follows from Fig. 6.4a that due to their angular dependence, screening effects may vary by a factor as large as 2 in the present case. Hence, the approximate formula

(6.10) should be used with care as it is not dependent on the emission angle. Close inspection of Fig. 6.4a indicates that the angular variation causes a displacement of the Z_p^{eff} curve along the x axis that could be accounted for by an adjustment of the c parameter.

More insight into the angular dependence of the screening effects is obtained from the elastic scattering model presented in Sect. 3.1.2. As in (5.4) the elastic scattering cross section is replaced by the Rutherford cross section. To account for the projectile electrons, we use the screened projectile charge $Z_p^{eff}(K_e')$ that depends on the momentum transfer K_e' of the scattered target electron. From first-order perturbation theory [Messiah (1962)], one obtains for the effective charge experienced by an elastically scattered electron, in accordance with (6.7):

$$Z_p^{eff}(K_e') = Z_p - S_p(K_e'), \qquad (6.11)$$

where S_p is the screening function from (6.6). Recall that primed quantities refer to the projectile frame of reference.

The Rutherford formula (3.2) contains the well-known expression for the momentum transfer:

$$K_e' = 2v_e' \sin \frac{\theta_s'}{2}, \qquad (6.12)$$

where $v_e' = \sqrt{2\epsilon'}$ is the incident electron velocity. The scattering angle θ_s' is equal to the electron observation angle θ' modified by the angular spread θ_s' of the incident electron beam, as given by (3.15).

We recall that the present treatment of screening is based on the analytic formula (5.4) used in conjunction with (6.11). The latter formula describes energy and angular dependence of the effective nuclear charge. Results from (6.11) are shown in Fig. 6.4b which refers to collisions of 0.5 MeV/u He$^+$ on H. The electron observation angle ranges from 30° to 150°. As discussed above, the results for 0° and 180° (not shown) are less accurate, however, they are free of singularities. In Fig. 6.3 the results of the elastic scattering model for 60° are compared with those for the other models. It is seen that the analytic formula yields an effective charge Z_p^{eff} that is very close to that of the Born approximation. This gives confidence that the essential features of the screening effects can be described within the framework of the elastic scattering model.

The present models describe screening in momentum space. Alternative attempts were made to evaluate screening in configuration space. In this case further approximations are needed to obtain analytic expressions. Toburen et al. (1981) put forward a simple screening model where the effective nuclear charge of the projectile is obtained by integration of the spatial distribution of the passive electrons. The integration limits are given by a sphere whose radius is equal to the internuclear distance. Outer screening is neglected. The charge cloud of the active electron is assumed to have negligible extension. When compared to experiment, the adiabatic radius is set equal to the internuclear distance. In spite of the various approximations, the model calculations were found to be in reasonable agreement with experiment [*Toburen* et al. (1981)]. Recently, improved screening models have been worked out in configuration space by *Montenegro* et al. (1992) and *Ricz* et al. (1993). These models, however, do not provide analytic solutions and they have not yet been applied to double differential cross sections.

Finally, we note that little information exists about angular dependencies of screening effects. *Kövér* et al. (1982) have studied the projectile screening for 0.8- and 1-MeV/u H_2^+ and He^+ as a function of the electron observation angle. Similar experiments were performed by *DuBois* et al. (1994). A detailed analysis of the angular dependence of the screening was hampered due to the presence of the electron-loss peak whose intensity varies strongly with angle. In view of the previous results it would be beneficial to perform further studies of screening in electron emission processes. Adequate treatment of electronic screening is required for an accurate evaluation of ionization cross sections by dressed projectiles. Also, higher-order effects of the mean field produced by the passive electron need to be taken into account as discussed in the following section.

6.3 Binary Encounter Electron Emission

As outlined in Sect. 3.7.2,5.1, binary collisions between projectiles and free electrons are classic examples of single-center two-body interactions leading to target ionization. Traditionally, theories treating these binary interactions for fully stripped ion impact predict that the cross sections scale quadratically with the projectile nuclear charge (5.4). For partially stripped ion impact, it was shown in the previous Section that screening of the nuclear charge should play a minor role for high-energy electron emission and, again, the BE cross sections are expected to scale quadratically with the projectile nuclear charge. Early data by *Stolterfoht* et al. (1974) appeared to verify these predictions, Fig. 6.1. However, recent studies involving heavy ions colliding with atoms have demonstrated deviations from these expectations and have led to a renewed interest in what was assumed to be a well-understood ionization mechanism.

6.3.1 Enhancement of the Binary Encounter Peak

This renewed interest began a few years ago when a group at Kansas State University initiated a detailed investigation of binary-encounter electron emission. Initially, the zero-degree BE intensity for fully stripped ion impact, $Z_p \leq 10$, was investigated, [*Lee* et al. (1990)], and Z_p^2 scaling was confirmed, in accordance with expectations. [See Fig. 5.3 and the associated discussion.] Shortly thereafter, *Richard* et al. (1990) extended these studies to dressed fluorine ion impact and found the unexpected result that the zero-degree BE

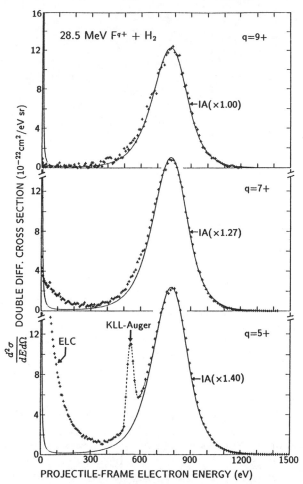

Fig. 6.5. Binary-encounter electron spectra transformed in the projectile frame for 1.5-MeV/u $F^{q+} + H_2$ collisions with q= 9, 7, 5. The *points* refer to the experimental data. The *full curves* refer to impulse approximations (IA) calculations for bare F^{9+} projectiles normalized to the experimental data by multiplying with the numbers in parenthesis. From the work by *Richard* et al. (1990)

cross sections increased as electrons were added to the projectile. This feature is illustrated in Fig. 6.5 for 1.5 MeV/u F^{q+} ions impacting on a molecular hydrogen target.

The theoretical calculations performed using the elastic scattering model (Sect. 3.1.2), labeled IA, are in excellent agreement with the experimental data for fully stripped ion impact. However, for partially stripped ion impact, they underestimate the magnitude of the binary-encounter cross section which steadily increases as the net charge of the projectile decreases. Later experiments, performed by *Hidmi* et al. (1993), verified this anomalous scal-

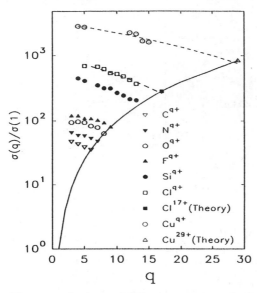

Fig. 6.6. Scaling of BE cross sections for bare and dressed ion impact. Impact energies are 1-MeV/u except for copper projectiles where 0.5-MeV/u was used. The *open square* and *circle* indicated by *arrows* are theoretical values for fully stripped chlorine and copper ions. The *solid curve*, drawn through the fully stripped ion data, indicates Z_p^2 scaling of proton impact BE cross sections. Data are from *Hidmi* et al. (1993)

ing behavior for additional collision systems and for projectile charge states up to 29 (Fig. 6.6).

Considerable theoretical interest was generated by these observations. For small observation angles an inverse q dependence of the BE cross section, i.e., an enhancement of the electron-ion scattering, has different explanations which are more or less related. Generally, the enhancement results because the mean field of the bound projectile electrons modifies the Coulomb potential of the projectile. When a target electron penetrates deeply into the resulting non-Coulombic potential, it scatters more strongly than it would in a pure Coulomb field of a bare ion [*Reinhold* et al. (1990a), *Shingal* et al. (1990), *Schultz* and *Olson* (1991)]. One interpretation is that at small distances the bound electron increases the slope of the Coulomb potential so that the force (equal to the derivative of the potential) on the scattered electron increases.

Thus, the presence of the bound projectile electrons enhances the classical deflection function for the incident electron. This modification can lead to deflections exceeding 180° for small impact parameter collisions [*Reinhold* et al. (1991), *Schultz* and *Olson* (1991)]. Remember that 180° scattering in the projectile frame is viewed as 0° in the laboratory frame. When this occurs, the BE cross sections are enhanced with respect to those for fully stripped ion impact.

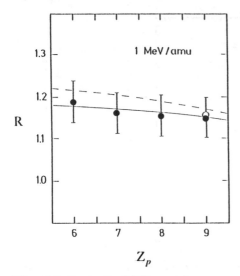

Fig. 6.7. Ratio R of differential cross sections for the BE maximun as a function of the projectile nuclear charge. *Open circle*: experimental data from *Richard* et al. (1990), *solid circles*: experimental data from *González* et al. (1992), *solid line* and *dashed lines*: distorted wave calculations (see text). From *Ponce* et al. (1993)

Electron exchange effects can further enhance the zero degree BE emission, as was demonstrated by *Taulbjerg* (1990b) and *Bhalla* and *Shingal* (1991). In Fig. 6.7 the ratio

$$R(\theta' = \pi) = \frac{d\sigma(q)}{d\Omega} \bigg/ \frac{d\sigma(Z_p)}{d\Omega} = \left(\frac{1}{4}\left|f^+\right|^2 + \frac{3}{4}\left|f^-\right|^2\right) \bigg/ \left|f_{Z_p}^c\right|^2 \quad (6.13)$$

is plotted for impact of 1 MeV/u monoelectronic projectiles on H_2 targets [*Ponce* et al. (1993)]. The data are for electrons with energies corresponding to the maximum of the BE peak. In (6.13) θ' is the electron scattering angle in the projectile system and $d\sigma(q)/d\Omega$ is the differential cross section for scattering of an electron in the field of a monoelectronic projectile of net charge q. Exchange contributions are taken into account and the differential cross sections are averaged over singlet and triplet states. Moreover, $\left|f_{Z_p}^c\right|^2$ is the Rutherford cross section for elastic scattering of the electron in a Coulomb field of charge Z_p. More information about the theoretical method is given in Sect. 3.7.2.

Thus, under certain conditions for dressed ion impact, the BE electron emission cross sections increase as the net charge on the projectile decreases. However, under other conditions, e.g., other impact energies or emission angles, theoretical studies by *Schultz* and *Olson* (1991) predicted that this enhanced behavior may not occur. Using CTMC calculations, they found that the BE cross sections for dressed ion impact can be enhanced, suppressed, or unchanged with respect to the cross sections for bare ion impact.

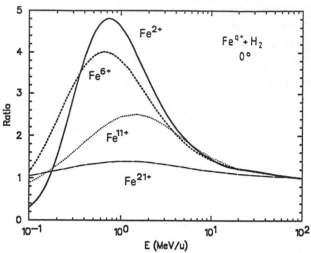

Fig. 6.8. CTMC predictions of binary-encounter electron emission at 0° resulting from Fe^{q+} impact. The ratios, partially-stripped to fully stripped ion impact, are shown as a function of projectile charge state and impact energy. From *Schultz* and *Olson* (1991)

These predicted features can be observed from the theoretical results given in Fig. 6.8. The figure shows cross section ratios for BE electron emission at 0° resulting from partially stripped iron projectiles with net charge states from 2 to 21. These cross sections were divided by those for fully stripped iron ions, Fe^{26+}, thus yielding the ratios shown. Note that for a range of collision energies the BE cross sections for partially stripped ion impact are larger than those for fully stripped ion impact and that the cross sections decrease as the net charge increases. Also note that the enhancement of the BE cross sections disappears at very high impact energies where the cross sections are independent of the projectile charge state.

Recent data obtained at Kansas State University for 1-MeV/u F^{q+} ions impacting on H_2 [*Liao* et al. (1994), *Richard* (1996)] shown in Fig. 6.9, confirm these predicted features. In qualitative accordance with the theoretical data in Fig. 6.8, the enhancement of the binary-encounter electron emission seen for 0° diminishes at 10°. At an angles of 40° the BE peak intensity shows the usual behavior due to screening, i.e., a decrease with decreasing projectile charge state. This charge-state dependence is in agreement with the earlier binary-encounter results at 25°, previously shown in Fig. 6.1.

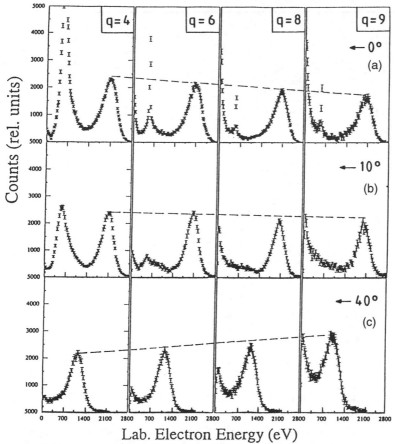

Fig. 6.9a–c. Electron emission spectra for $1\,\mathrm{MeV/u}$ F^{q+} impact on H_2 for the charge states $q = 4$, 6, 8 and 9. The peak centered near $2100\,\mathrm{eV}$ is due to BE emission. The data are observed at $0°$ (**a**), $10°$ (**b**), and $40°$ (**c**). Note the unusual decrease in the BE intensity with increasing projectile charge state at $0°$. Data from *Liao* et al. (1994) and *Richard* et al. (1996)

6.3.2 Diffraction Effects

At approximately the same time that the scaling anomaly was being reported, another unexpected feature in the binary-encounter emission was found by researchers at the University of Frankfurt [*Kelbch* et al. (1989a,b)]. This anomaly is illustrated in Fig. 6.10. While investigating energetic collisions between heavy particles, e.g., $1.4\,\mathrm{MeV/u}$ $U^{32,33+}$ on rare gases, the binary peak was observed to suddenly disappear and then to reappear at a different emission energy when the observation angle was varied. Further investigations by the Frankfurt and Kansas State University groups [*Wolff* et al. (1992, 1993, 1995), *Wolf* et al. (1993), *Shinpaugh* et al. (1993), *Liao* et al. (1992)] have demonstrated that this is a rather general feature. Also, it

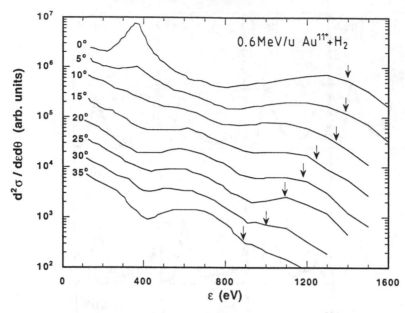

Fig. 6.10. DDCS for electron emission in $0.6\,\mathrm{MeV/u}$ Au^{11+} + H_2 collisions. Arrows indicate the expected location of the binary encounter peak as predicted by simple two-body kinematics. Note that the data from *Wolff* et al. (1993) have been arbitrarily scaled at each angle for purposes of display

was confirmed that this structure is associated with the non-Coulombic field of dressed projectiles.

The oscillatory structure in the BE electron spectra can be understood in the framework of a free-electron model. It is recalled that the emission of the binary-encounter electrons is equivalent to a scattering between a target electron and the projectile ion. For certain charge state ions, target electrons elastically scattering from a non-Coulombic field of the projectile have pronounced minima in the elastic scattering cross sections [*Reinhold* et al. (1991), *Schultz* and *Olson* (1991)]. As evident in the early work by *Ramsauer* (1921) and *Townsend* and *Bailey* (1922), differential cross sections for electron-atom scattering exhibit oscillations due to interference effects. In a quantum-mechanical wave picture these oscillations can be attributed to diffraction effects. Thus, the BE emission may have minima and maxima at specific emission angles.

For diffraction effects to be observable in the BE emission, certain conditions must be fulfilled. First, as noted in Sect. 2.2, the target electron has a distribution of momenta. This means that diffraction effects will only be distinguishable when the minima in the elastic scattering cross section are not washed out by the convolution over the electron momentum distribution, i.e., diffraction effects will be strongest when the target electron momentum distribution is narrow, as demonstrated by *Wolff* et al. (1993).

Second, minima in the elastic scattering cross sections indicate that the cross sections are dominated by only a few partial waves. Few partial waves imply that the impact energy is low. Thus, the diffraction effects in the BE emission should occur at lower-impact energies and should gradually disappear at higher impact energies. These expectations were investigated by *Shinpaugh* et al. (1993) who studied BE emission for 0.6 to 3.6 MeV/u I^{23+} impact on He and Ar. They found that although diffraction effects tended to disappear with increasing impact energy the effects were still observable at the highest energy studied.

In summary, binary-encounter cross sections exhibit diffraction effects under conditions where the cross sections for electrons elastically scattering from the projectile contain localized minima. These effects will be strongest when the target momentum distribution is smallest. For more detailed information the reader is referred to the recent review by *Lucas* et al. (1997). The authors discuss also diffraction effects observed for the electron loss peak. This peak corresponds to the binary encounter peak when the collision system is reversed, i.e, when the collision partners are interchanged. The electron loss peak will be discussed in the following section.

6.4 Electron Emission from the Projectile

When studying electron emission from the projectile, i.e., electron loss to the continuum (ELC), it is important to realize that the ionization mechanisms for the projectile are in principle the same as those for the target. If the roles of target and projectile are reversed, the essential properties of the ionization process are unchanged. Thus, the substantial points of the discussion of target ionization in the previous sections are also valid for electron loss.

A few differences, however, are to be taken into account. First, with respect to projectile ionization, the screening effects for the neutral target are significant. The long-range dipole interaction, characteristic for incident ions are not present for projectile ionization. Second, dielectronic processes will be relatively important since the neutral target is dressed with a maximum number of electrons whose binding is relatively small. Lastly, the electrons ejected from the moving projectile are strongly influenced by kinematic effects. In the following, the properties of the electron loss mechanism shall be discussed in conjunction with theory and experiment.

6.4.1 Theoretical Considerations

In electron emission spectra, the electron loss peak has independently been discovered by *Burch* et al. (1973) and *Wilson* and *Toburen* (1973). In these early investigations, adequate interpretations for the mechanisms operative in the electron-loss mechanisms were given. The electrons on the incident ion were assumed to constitute a merged beam of quasi-free particles traveling with the velocity v of the projectile. *Wilson* and *Toburen* (1973) integrated the electron-loss peak and compared the results with scattering cross sections of free electrons. *Burch* et al. (1973) also took into account the momentum distribution $f(p')$ resulting from the atomic binding where p' is the momentum of the electron in the projectile frame. An electron with the momentum $p = v + p'$ is then considered to scatter elastically from the target atom in the laboratory frame of reference.

The theoretical model by *Burch* et al. (1973) is similar to the binary-encounter theory treated in Sect. 3.1.3. It should be noted, however, that the present case refers to a reversed collision system where the heavy particle (target) is at rest. Thus, the double-differential cross section for electron emission is obtained by integrating over the momentum direction $\hat{p}'$ of the bound electron,

$$\frac{d\sigma}{d\Omega\, d\epsilon} = \int\limits_{4\pi} \frac{d\sigma(\epsilon, \theta, \hat{p})}{d\Omega}\, f(p')\, \frac{dp}{d\epsilon}\, d\hat{p}', \tag{6.14}$$

where $f(p') = |\varphi(p')|^2$ is the velocity distribution of the bound electron obtained from its momentum wave function. Since the electron is elastically scattered, it follows that $\epsilon = p^2/2$ and, hence, $d\epsilon/dp = p$. For the elastic scattering cross section, *Burch* et al. (1973) used the Rutherford formula (3.3) modified by exponential screening.

In Fig. 6.11 theoretical results are compared with electron spectra measured in 23- to 41-MeV O^{4+} + Ar collisions by *Burch* et al. (1973). The spectra show the pronounced electron-loss maximum that is shifted with projectile energy. Reasonable agreement between the theoretical and experimental results is obtained showing that the electron loss mechanism is adequately described by the present model. The peak position is predicted to be close to the reduced energy $T = v^2/2$. It is seen, however, that the EL peak is shifted to lower energies in comparison with the T value indicated by arrows. This energy shift shall not be further discussed here, since it is analogous to that analyzed previously for the binary-encounter peak.

The treatment by *Burch* et al. (1973) is similar to the elastic scattering model presented in Sect. 3.1.2. In that model, an integration is performed over the transverse momentum of the bound electron (3.14). When this formalism is applied to the reversed collision system, primed quantities are to be exchanged by the corresponding unprimed quantities (and vice versa) and the frame transformation factor $\sqrt{\epsilon/\epsilon'}$ is to be left out. Hence, from (3.11) it follows that

$$\frac{d\sigma}{d\Omega\, d\epsilon} = \frac{d\sigma(\epsilon, \theta)}{d\Omega}\, \frac{dp_z}{d\epsilon}\, J(p'_z). \tag{6.15}$$

We recall that $J(p'_z)$ is the Compton profile and $dp_z/d\epsilon = (v + p'_z)^{-1} \approx v^{-1}$ for fast collisions. Since no integration is required, the elastic scattering model can easily be evaluated. Its well-known disadvantage of producing inaccurate

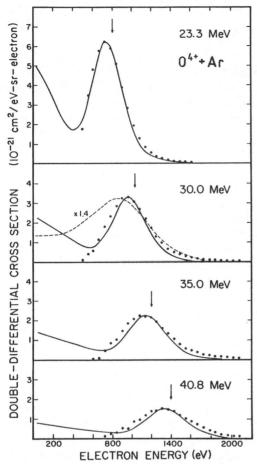

Fig. 6.11. Experimental and theoretical cross sections for electron emission at 90° in O^{4+} + Ar collisions. The projectile energy is varies as indicated. The experimental data points are normalized to the theoretical curves given by *solid lines*. The *dashed line* results from a 1s velocity distribution with a 2s binding energy. From *Burch* et al. (1973)

results at small scattering angles is less severe for projectile ionization, as the elastic scattering cross section for neutral (target) atoms does not imply a singularity at $\theta = 0$. The first analysis of electron-loss maxima in terms of the Compton profile has been conducted by *Böckl* and *Bell* (1983).

Drepper and *Briggs* (1976) applied an alternative concept, i.e., that of first calculating projectile ionization in the projectile frame and then performing a frame transformation. In particular, they applied the Born approximation to evaluate the cross section $d\sigma'/d\Omega'\,d\epsilon'$ for electron loss in the projectile frame of reference. The corresponding cross section $d\sigma/d\Omega\,d\epsilon$ in the laboratory reference frame were obtained by the well-known expression:

$$\frac{d\sigma}{d\Omega\, d\epsilon} = \sqrt{\frac{\epsilon}{\epsilon'}} \frac{d\sigma'}{d\Omega'\, d\epsilon'},\tag{6.16}$$

where the transformation rules for energy and angle are given in Appendix B. Similar work was carried out by *Manson* and *Toburen* (1981) who compared theoretical results for the sum of electron loss and target ionization with experiment. Comparisons were also made by *DuBois* and *Manson* (1986, 1990) for individual target and projectile contributions which were separated by means of coincidence techniques. The good agreement between theory and experiment shows that the Born approximation is suitable to describe the experimental results for collisions of 0.5-MeV/u He$^+$ on He.

Drepper and *Briggs* (1976) pointed out important features of the electron-loss process. In the projectile frame, electrons from soft- and binary-collisions are observed as distinct structures. However, the transformation from the projectile to the target reference frame shifts the soft- and binary-collision peaks to a common energy, so both are found in the electron-loss peak. Soft collision electrons, ejected at nearly zero energy in the projectile rest frame, are found at 0° and travel with a velocity nearly equal to the projectile velocity v. Again, in the projectile frame, binary collision electrons are primarily ejected in the direction of the "incident" target atom and have the velocity $2v$. In the laboratory frame of reference these electrons are ejected at 180° and have the velocity $v = 2v - v$. Hence, in the laboratory frame, soft-collision and binary-collision electrons have the same energies. They are, however, ejected in opposite directions. In the laboratory frame, soft-collision electrons are found at 0°, whereas binary-collision electrons are found at 180°.

In the theoretical description of the electron-loss mechanisms, the previous conclusions about soft- and binary collisions remain applicable. Soft-collision electrons ejected at forward angles are well-described using quantum-mechanical methods. Hence, it is favorable to use (6.16) in conjunction with the Born approximation, which also allows the treatment of screening effects in first-order. Binary collisions, associated with large observation angles can be described by two-body theories such as the elastic scattering model. As discussed previously, the model allows for the inclusion of higher-order effects in the screening of the target atom. In the past, considerable effort has been performed to describe the electron-loss peak. Early studies are discussed in the work by *Breinig* et al. (1982), *Prost* et al. (1982), *Burgdörfer* et al. (1983), and *Schneider* et al. (1983). More recent work was reported by *DuBois* and *Manson* (1990) and *Wang* et al. (1992).

6.4.2 Experimental Studies

In the following, the attention is focused on the experimental effort concerning the electron-loss mechanism. The experiments by *Burch* et al. (1973) demonstrated that the electron-loss peak appears as a pronounced structure in the electron spectra. Nevertheless, it is superimposed on a continuous background whose intensity is significant. The analysis by *Manson* and *Toburen* (1981) has shown that this background is primarily due to target ionization. Hence, certain effort is required to separate the ELP from other contributions. In general, the peak-to-background ratio increases with increasing projectile energy. Hence, it is favorable to study the electron-loss peak at high projectile energies. Backgrounds are particularly troublesome at backward angles of the ejected electrons, since the elastic scattering cross section decreases strongly with increasing angle. Therefore, we shall focus our attention to high incident energies when backward angles are considered.

However, at forward angles, in particular at 0°, the electron-loss peak can be observed for relatively low-collision energies since this peak forms a cusp-like structure similar to that for the electron capture to the continuum process. Nevertheless, at forward angles, the subtraction of the continuous background still represents a severe problem. In addition, the shape of the 0° electron-loss peak depends on various experimental parameters, such as the acceptance angle of the spectrometer, so that the analysis of the EL cusp is often limited.

Typical electron-loss spectra, observed at 0°, are given in the upper part of Fig. 6.12. The ELC data were measured by *Kövér* et al. (1986) in collisions of 0.5-, 1-, and 2-MeV He$^+$ on He. They are compared with corresponding ECC spectra obtained using He^{2+} impact and given in the lower part of the figure. The distinguishing feature of the ELC peak is its rather symmetric profile. This is in contrast to the asymmetric peak profile observed for the ECC peak (refer also to the discussion in the preceding section). For the electron loss mechanism, the peak shape is essentially determined by the kinematic factor $\sqrt{\epsilon/\epsilon'}$ in (6.16) which primarily explains the symmetric cusp profile.

The shape of the ELC peak has been subject of intense discussions in the literature as it yields information about electron production mechanism at threshold. *Drepper* and *Briggs* (1976) and *Briggs* and *Day* (1980) have shown within the framework of the Born approximation that for hydrogenic wave functions and for negligibly small electron energies, the non-zero terms in the Legendre polynomial expansion (5.6) are even. In this case, a symmetric cusp is produced. Using methods of group theory associated with the pure Coulomb potential, *Burgdörfer* et al. (1983) and *Burgdörfer* (1986) performed elaborate studies of the shape and asymmetry of ELC cusps. Comparison between theory and experiment was primarily performed for ions with an initial 1s state. For an initial 2p (m=0) state, *Burgdörfer* (1983) predicted an inverted cusp which still consists of a pronounced maximum but has a dip at the location where one expects the cusp. It should be recalled that the theory by *Burgdörfer* and collaborators (1983,1984,1986) is based on hydrogenic wave functions.

To analyze the ELC peak profile in detail, it is instructive to transform the data back into the projectile frame of reference. Thus, it will become evident that the symmetry in the ELC peak, observed in the laboratory frame, corresponds to a forward-backward symmetry of the slow electron

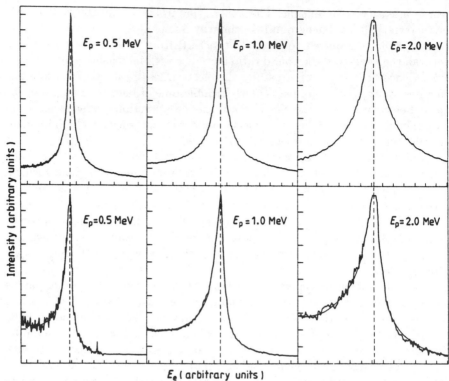

Fig. 6.12. Comparison of cusp spectra due to electron loss to the continuum (upper row) in 0.5-, 1.0-, 2.0-MeV He^+ + He collisions with corresponding spectra due to electron capture to the continuum (*lower row*). From *Kövér* et al. (1986)

emission in the projectile frame. The symmetry of soft-collision electrons has already been discussed in detail in Sect. 5.2.3 treating also non-hydrogenic wave functions. It has been shown that various effects cause asymmetries in the emission of soft-collision electrons. However, at high incident energies and low projectile charge state, electron emission becomes more and more symmetric. In particular, in a reversed collision system, it is likely that for electron ejection by neutral (target) particles two-center effects are negligible. This is an essential reason for the observed symmetry of the ELC peak.

It is recalled, however, that the absence of two-center effects is not a sufficient condition for the isotropy of the slow electron emission. In accordance with the discussion in Sect. 5.2.3 for low impact energies, one would expect an anisotropic angular distribution (Fig. 5.13) and, hence, certain asymmetries in the EL peak profile. In particular, as pointed out before, the angular distributions may be subject to significant variations as the electron energy tends to threshold. This effect, predicted theoretically, has not yet been confirmed experimentally. It is important to again note that the analysis of the peak profile provides information about electron emission from the projectile with very low energies. (By target electron spectroscopy these low-energy electrons are usually inaccessible due to well-known instrumental problems). Precision measurements of the EL peak such as that by *Kövér* et al. (1986) would be useful to verify the asymmetric angular distributions of the ejected electrons.

However, it should be kept in mind that apart from small asymmetries, the 0° cusp is essentially determined by the kinematic factor $\sqrt{\epsilon/\epsilon'}$. To obtain more information about the atomic properties of the active electrons, measurements at angles other than 0° are required. For larger angles the electron-loss peak is primarily determined by the Compton profile of the bound projectile electron (6.15). When the projectile carries electrons in different shells, the Compton profile is usually composed of contributions from these shells (or subshells). Information about individual shell contributions may be obtained from separate calculations of the Compton profile.

Examples for the Compton profile analysis of electron-loss peaks are given in Fig. 6.13. The experimental data are obtained for collisions of 120-MeV Ne^{5+} and 200-MeV Ne^{7+} on Ne at an observation angle of 30°, *Schneider* et al. (1983). The theoretical results for individual shells and subshells were evaluated using a binary-encounter formula similar to that in (6.14). The theoretical results clearly show the different subshell contributions. The Ne^{5+} projectile has one 2p and two 2s electrons. Hence, we may understand that the 2s contribution is more intense than that for 2p, Fig. 6.13a. It is interesting to note that the 1s contribution is found to be significant. From Fig. 6.13b we see that for Ne^{7+} the 1s contribution becomes as important as that for 2s. Since the 1s curve is strongly shifted in energy, the 1s contribution is detected as a low-energy shoulder in the electron spectrum. Thus, in this example for the electron-loss process 1s electrons (bound by about 1 keV) play a similar role as the much weaker bound 2s electrons.

The angular dependence of the projectile electron emission is determined by the corresponding elastic scattering cross section which, in turn, depends strongly on the mean field of the passive target electrons. In particular, screening of the target nucleus is expected to play a significant role. *Schneider* et al. (1983) measured angular distributions of the projectile electrons for target atoms whose nuclear charge was strongly varied. The results for collisions of 105-MeV Ar^{5+} on Ne, Kr, and Xe are given in Fig. 6.14, where spectra for the observation angles of 90° and 150° are compared. It is seen that in the case of Ne the spectral intensity for 150° is much lower that for 90°. Such decrease is indeed expected as the Rutherford scattering cross section decreases strongly with increasing angle. However, Fig. 6.14 shows that the angular dependence is significantly reduced for Kr and is diminishingly small for Xe.

This finding may be explained by the mean field effects of the passive target electrons. Active electrons that are backward scattered penetrate deeply into the target atom. Therefore, the 150° electrons feel the increase of the target nuclear charge more strongly than the 90° electrons. Thus, the scatter-

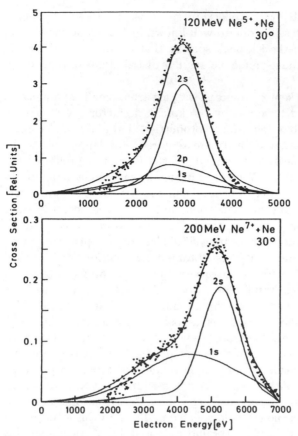

Fig. 6.13. Subshell composition of electron loss peaks for the collision systems 120-MeV Ne^{5+}+ Ne and 200-MeV Ne^{7+}+ Ne. *Solid lines* represent the theoretical results and *points* refer to experiments. The continuous background is subtracted. The experimental data are normalized to theory. Electron observation angle is 30°. From *Schneider* et al. (1983)

ing at 150° increases in relation to that of 90° as the target atom is changed from Ne to Xe. To estimate whether screening effects may be responsible for the observed variations of the 150° to 90° intensity ratio, we applied the first-order formula (6.11) in connection with (6.12). It was readily found that the present variation of the 150° to 90° intensity ratio is significantly larger than predicted by the first-order formula. Hence, higher-order effects of the mean field are expected to play an important role.

The need for a higher-order treatment has previously been recognized by *Jakubaßa* (1980) and *Hartley* and *Walters* (1987) after it was shown by *Walters* (1975) that the Born approximation is insufficient to describe the electron-loss event at backward angles. An extended theoretical approach, based on an impulse approximation with an accurate off-shell transition amp-

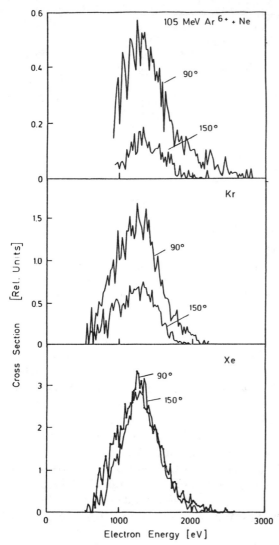

Fig. 6.14. Experimental cross sections for electron loss observed at 90° and 150° in a collision of 105-MeV Ar^{6+} on the target gases Ne, Kr, and Xe. The continuous background is subtracted. Structures in the spectra are due to statistical fluctuations. From *Schneider* et al. (1983)

litude, has recently been presented by *Wang* et al. (1991,1992) to describe electron loss at backward angles. Figure 6.15 shows theoretical results in comparison with experiments by *Heil* (1991) who measured EL at large backward observation angles. Surprisingly, the experimental data depict an electron intensity that increases by a factor of ~2 as the angle increases from 130° to 150°. The experimental results are in good agreement with the theory by

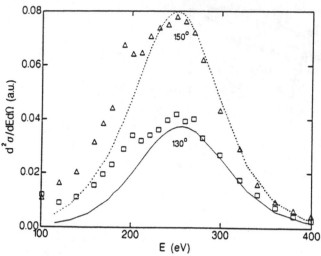

Fig. 6.15. Electron-loss cross sections for collisions of 0.5-MeV H° impact on Ar. Observation angles are 130° and 150° as indicated. The experiments (*points*) are from *Heil* (1991) and the theoretical results (*lines*) are from *Wang* et al. (1991) using an off-shell impulse approximation

Wang et al. (1991), indicating that the increase of the spectral intensity is real. It is evident that this increase cannot be accounted for in first-order. Rather, the present data provide clear evidence for higher-order effects of the mean field created by passive target electrons.

Higher-order phenomena can be taken into account by means of the elastic scattering model used in conjunction with accurate scattering cross sections in (6.15). The method has been discussed in detail in Sect. 6.3 devoted to the binary-encounter mechanism. It is recalled that the electron loss at backward angles results from binary-encounter electron emission from the target. The present cases show that higher-order phenomena, similar to those for the binary-encounter peak, may be observed for projectile electrons ejected at backward angles.

Also, it is expected that the electron loss peak shows diffraction effects similar as those discussed for the binary encounter maximum (Sect. 6.3.2). Looking back at (6.15) it is noted that the angular distribution of the electron loss peak is proportional to the scattering cross section for free electrons. As mentioned in Sect. 6.3.2 the differential electron scattering cross section exhibits diffraction pattern. They are due to a limited number of outgoing partial waves producing interference effects known since the early work by *Ramsauer* (1921). The Compton profile does not depend on the scattering angle. Hence, the diffraction pattern of the electron scattering may directly be visible in the angular distribution of the electron loss peak.

Diffraction phenomena have been observed in the electron loss peak by *Kuzel* et al. (1993b). Since then, these phenomena have been studied for

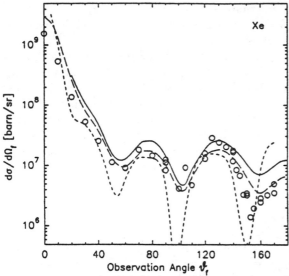

Fig. 6.16. Single differential cross sections for electron loss in collisions of 0.5-MeV H° on Xe as a function of the observation angle. The experimental data (*points*) are compared with results from a quantum mechanical impulse approximation (*dashed* and *solid lines*) which do and do not include dielectronic processes, respectively. The short *short-dashed* line is due to calculations for free electrons. From *Kuzel* et al. (1994)

various collision systems including rare gas targets [*Lucas* et al. (1997)]. In accordance with electron-atom scattering it was found that the oscillations of the electron-loss peak intensity increase with the complexity of the target. Hence, diffraction pattern are particular pronounced for Kr and Xe. An example for the system 0.5 MeV H° + Xe is given in Fig. 6.16 which clearly shows the oscillatory behavior of the integrated electron loss peak intensity [*Kuzel* et al. (1994)]. The experimental data are in good agreement with theoretical results obtained by means of quantum-mechanical impulse approximation without and with inclusion of dielectronic effects (see next section). More information about diffraction effects in electron loss and their interpretations by means of free-electron models may be obtained from the review by *Lucas* et al. (1997).

6.5 Dielectronic Processes

We now turn our attention to dielectronic processes and concentrate on (i) the importance of these processes in various dressed ion-atom collision systems and (ii) how dielectronic processes influence the total and differential ionization cross sections. Two-center dielectronic, $e - e$, interactions result in target and projectile electrons being excited to discrete or continuum states whereas screened nuclear interactions, $e - n$, are monoelectronic processes where either the projectile or the target is excited or ionized while the collision partner remains in the ground state.

For target ionization, as outlined by (6.1), dielectronic processes are reactions (6) and (8), i.e., interactions where the target is ionized and the projectile is excited to a discrete state or to the continuum. The monoelectronic channel, reaction (2), corresponds to target ionization with the projectile remaining in the ground state. Note that experimental techniques employing coincidence measurements between projectile and target ions are not sensitive to excitation of either the target or the projectile. Thus reaction (6) is falsely included with reaction (2) and the relative importance of dielectronic processes can be underestimated.

6.5.1 Threshold Behaviour

Dielectronic processes involve inelastic collisions between bound electrons. Hence, one identifying characteristic is a poorly defined threshold which the projectile energy, E_p, must exceed. This threshold energy is given by

$$E_p = M_p \, \Delta E, \tag{6.17}$$

where M_p is the projectile mass (in electron mass units) and ΔE is the minimum energy required for the inelastic process, e.g., it is the energy required to excite/ionize both of the electrons involved. If the incident electron was free, this inelastic scattering process would have a sharp impact energy threshold at the projectile energy E_p. Since the electron is bound, it has a distribution of momenta and the threshold behavior is smeared out. The threshold behavior for excitation by a dielectronic process was first observed by *Zouros* et al. (1989).

A second characteristic is that electrons emitted via dielectronic processes have velocities ranging from approximately 0 to v in magnitude. Again, if dielectronic processes involved interactions between two free electrons, there would be a sharply defined upper limit at v. When bound electrons are involved, the upper limit is smeared out and electrons with velocities larger than v are produced, but with rapidly decreasing probabilities. This means that first-order dielectronic processes cannot produce significant numbers of electrons with velocities much larger than the projectile velocity.

As stated, dielectronic ionization processes result in electrons being ejected from both collision partners. However, mutual ionization of the collision partners can also result from higher-order screened nuclear interactions, e.g., where a target electron is ionized via an interaction with the screened projectile nucleus and, independently, the projectile electron is ionized via an interaction with the screened target nucleus. These second-order processes should constitute a relatively small component of the differential electron emission at high impact energies because they have a v^{-4} dependence whereas the first-order dielectronic processes behave as v^{-2}. On the other hand, at lower impact energies, higher-order $e - n$ processes can become dominant and may disguise any sharp threshold behavior associated with dielectronic processes.

Fortunately, a signature of these second-order $e - n$ processes exists in the differential electron emission spectra. Since they involve collisions with massive nuclei, the emitted electrons have velocities ranging from 0 and $2v$ rather than from 0 to v as is the case for first-order dielectronic processes. Thus a clear and definitive signature of second-order $e - n$ processes is mutual ionization of the collision partners leading to the emission of electrons having velocities significantly larger than v.

6.5.2 Total Cross Sections

The first inclusion of two-center dielectronic processes in a theoretical treatment was approximately 40 years ago when *Bates* and *Griffing* (1955) applied the first Born approximation to collisions between hydrogen atoms. Quoting from their paper, the computational problem "though formidable in practice, is trivial in principle." More than a decade later, Bell and collaborators [*Bell* et al. (1969a,b)] extended these studies to dressed hydrogen and helium impact on helium. Their studies outlined the basic framework for performing first Born calculations of the $e - n$ and dielectronic interactions; the relevant formulae for differential cross sections were given in Sect. 3.7.2. Total cross sections are then obtained by integration over emission energies and angles. More recently, related frameworks have been discussed in the literature [*Gillespie* et al. (1978), *Briggs* and *Taulbjerg* (1978), *Gillespie* (1979), *Gillespie* and *Inokuti* (1980), *McGuire* et al. (1981), *Manson* and *Toburen* (1981) *Anholt* (1986), *DuBois* and *Manson* (1990), *Montenegro* et al. (1992, 1993a,b), *Kabachnik* (1993)].

Experimentally, it was nearly a decade after the initial theoretical studies before data became available for testing the predictions. *Horsdal-Pedersen* and *Larsen* (1979) used projectile ion-target ion coincidences to isolate the dielectronic interaction channel in energetic H-atom collisions. More recent studies applying Auger spectroscopy, *Zouros* et al. (1989), charge state analysis of the projectile beam [*Hülskötter* et al. (1990), *Shah* and *Gilbody,* (1991)] and coincidences with recoil ions [*Montenegro* et al. (1992), *Dörner* et al. (1994), *Wu* et al. (1994)] have provided evidence for the importance of two-center dielectronic processes. Moreover, electron spectroscopy experiments for atomic hydrogen [*DuBois* and *Kövèr* (1989)] and helium impact [*DuBois* and *Toburen* (1988), *Sanders* et al. (1995)] and for He$^+$ impact [*DuBois* (1989), *Montenegro* et al. (1992, 1993b)] permit further testing of the Born predictions.

Comparisons of experiment (dotted curves) and first Born theoretical predictions (solid curves) are shown for these light systems, e.g., H, He and He$^+$ impact on He, in Fig. 6.17. The comparison is for impact energies ranging from approximately 25 keV/u to 1 MeV/u. In Fig. 6.17a, cross sections for target ionization for $e - n$ interactions are given. Experimentally, projectile excitation processes due to channel (6), see (6.1), are included but theoretical

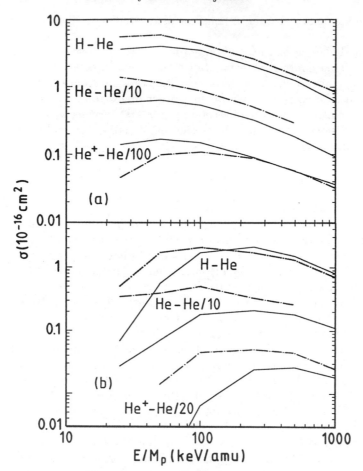

Fig. 6.17. Monoelectronic and dielectronic ionization cross sections for dressed hydrogen and helium particle impact on helium. The experimental data (*dotted curves*) are those of *DuBois* and *Kövèr* (1989), *DuBois* (1994), and *DuBois* (1987). The *solid curves* are B1 calculations by *Bell* et al. (1969a,b). Note that the experimental data include only single ionization processes whereas the theoretical data are shown for processes where both ionization and ionization plus excitation processes are considered. (**a**) shows cross sections for processes (2) and (6). (**b**) provides cross sections for the dielectronic process (8). For the definition of the processes see Sect. 6.1

predictions [*Bell* et al. (1969a,b)] indicate that these processes tend to be less important than the $e - n$ process leading to target ionization, channel (2). In Fig. 6.17b, cross sections are shown for the dielectronic process leading to ionization of both collision partners, channel (8). Note that the data for the different systems have been vertically displaced for purposes of comparison.

The general features of the screened nuclear and the dielectronic processes are apparent. For example, in Fig. 6.17b the poorly defined thresholds for the dielectronic processes are evident and the dielectronic processes are shown to be much weaker for ion impact than for atom impact. This is opposite to the situation for $e - n$ processes, shown in part Fig. 6.17a, which are stronger for ion impact. Comparing data for the $e - e$ and the $e - n$ processes shows that dielectronic processes are the dominant target ionization mechanism for energetic atom impact but are far less important for ion impact. However, we note that the behavior shown here is for a light target, e.g., helium, and is different than that found for heavier targets [Horsdal-Pedersen and Larsen (1979)].

The relative magnitudes of the screened nuclear and the dielectronic processes can be understood by referring to (3.209,210) in Sect. 3.7.1 and recalling that total cross sections are dominated by distant collisions leading to minimum momentum transfer. Thus, qualitative information about the importance of screened nuclear and dielectronic processes can be obtained by evaluating the terms under the integral at the lower limit of integration, i.e., as $K \to K_m$. Recall that $K \to K_m$ implies $F_{o' \, o'} \to 1$. Evaluating the respective bracketed terms for the $e - n$ channel predicts that $\sigma_H < \sigma_{He} \ll \sigma_{He^+}$; for the dielectronic channel, the prediction is that $\sigma_H \approx \sigma_{He^+} \approx \sigma_{He}/2$. When differences in the limits of integration for the various collision systems are taken into account, the expectations are consistent with the experimental data.

Quantitatively, as shown in Fig. 6.17, the first Born approximation does an excellent job in describing the screened nuclear interactions for high energy ion impact and a reasonable job for atom impact. For dielectronic interactions, the description is quite good for H + He collisions but tends to underestimate this channel for He$^+$ impact. At lower impact energies, increasing deviations between experiment and this first-order theory are seen.

Evidence exists that most of the discrepancy may be attributed to the importance of second-order, independent $e - n$ processes. This is demonstrated in Fig. 6.18 where experimental data for He$^+$–He collisions are compared to theoretical predictions [Montenegro et al. (1992, 1993b)], which include both first-order $e - e$ and second-order $e - n$ processes leading to projectile and target ionization. An independent-particle model which assumed uncorrelated ionization of the target and the projectile was used, e.g.,

$$\sigma_{e_T - n_P, e_P - n_T} = 4\pi \int_0^\infty db\, b P_P(b) P_T(b) \left[1 - P_T(b)\right]. \tag{6.18}$$

Montenegro et al. (1993b) argue that since the first-order dielectronic processes occur predominantly at large impact parameters while second-order $e - n$ processes are restricted to smaller impact parameters, overlap between the two processes is small and the probabilities may be combined incoherently. As seen, inclusion of second-order processes yields much improved agreement with the experimental data.

Equations (3.209,210) also demonstrate that the target ionization cross sections should scale linearly with the number of projectile electrons, N_P, for first-order dielectronic processes whereas a quadratic behavior depending on

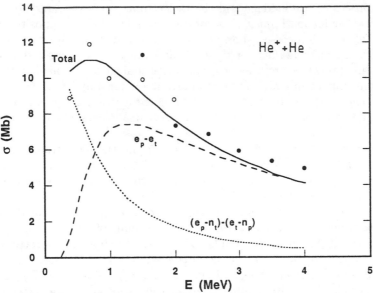

Fig. 6.18. Experimental and theoretical cross sections for dielectronic processes leading to ionization of both collision partners in He^+–He collision. The experimental data are from *DuBois* (1987), ○, and *Montenegro* et al. (1993), ●. The theoretical values are for first-order interactions between target and projectile electrons, dielectronic interactions, and for second-order $e - n$ interactions, i.e., independent interactions between a target electron and the screened projectile nucleus and a projectile electron with the screened target nucleus. The sum of the processes is also shown. Theoretical values are from *Montenegro* et al. (1993b)

the effective projectile nuclear charge is expected for first-order $e - n$ interactions. Thus, *DuBois* and *Manson* (1994) interpreted a linear N_P dependence of the total target ionization cross sections as evidence that two-center dielectronic processes dominate. But this interpretation is obviously incorrect, as demonstrated in Fig. 6.19. Here target ionization cross sections for both the dielectronic and the screened nuclear channels for ionization of helium by fast atom impact are plotted. It is seen that nearly linear dependencies on N_P result in both cases and thus their sum, i.e., the total target ionization cross section, also has a linear dependence.

More information about the role of the dielectronic processes have been achieved by *Sulik* et al. (1994) who studied the production of cusp electrons and $2s \rightarrow nl$ projectile excitation in collisions of 120-MeV Ne^{6+} with various target atoms. The results show a nearly linear dependence on the atomic number of the target atom ranging from $Z_t = 1 - 18$. From the theoretical analysis of the experimental data *Sulik* et al. (1994) concluded that for heavier target species the $2s \rightarrow nl$ excitation is governed by higher-order processes and the observation of a linear dependence should be considered as an accident.

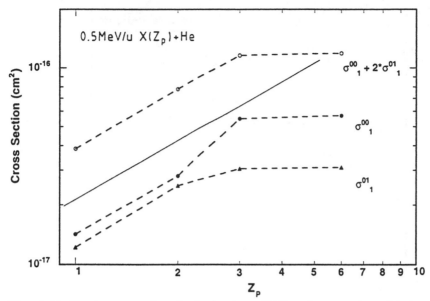

Fig. 6.19. Total cross sections for ionization of helium by neutral particle impact. The data are for the $e - e$ processes, $\sigma_1^{0\,1}$, e-n processes, $\sigma_1^{0\,0}$, and for the sum of these channels. *Dashed curves* through the experimental data [*DuBois* and *Kövèr* (1989), *DuBois* (1994), *Sanders* et al. (1994)] are to guide the eye. The *solid curve* demonstrates a linear dependence on the number of projectile electrons which is the expected behavior for $e - e$ processes

However, for the lighter targets, hydrogen and helium, the theoretical analysis predicts a dominant role of the dielectronic processes for the projectile excitation. This can directly be seen from the experimental spectra shown in Fig. 6.20 where, superimposed on the corresponding cusp spectrum, are Coster-Kronig lines due to $1s^2\,2p\,nl$ states produced by the transitions $2s \to nl$. Also seen is the surprising result that the spectrum excited by H_2 is larger than that produced by He. This inverse dependence on the target nuclear charge has been confirmed in the theoretical analysis, *Sulik* et al. (1995). It is explained by the fact that dielectronic processes are more probable when the target electron is less tightly bound, i.e., when the electron can favorably play an active role (Fig. 2.6). Hence, Fig. 6.20 provides direct evidence for dielectronic processes being important for the $2s \to nl$ excitation process. A similar conclusion has been drawn for the corresponding differential cross sections associated with the electron cusp, *Sulik* et al. (1995).

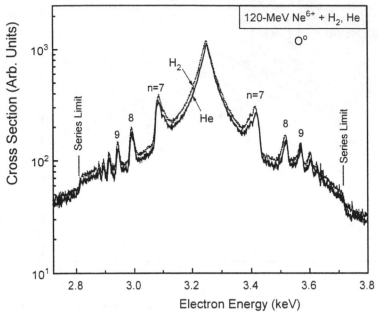

Fig. 6.20. Spectrum of Coster-Kronig electrons due to the initial projectile config-
urations $1s^2\,2p\,nl$ $(n = 7, 8, \ldots)$ superimposed on a $0°$ cusp electron spectra. The
collision systems are 120-MeV $Ne^{6+} + H_2$ and $Ne^{6+} + He$. Note the unexpected
result that the cross sections associated with the H_2 target are larger than those
attributed to He. From *Sulik* et al. (1995)

6.5.3 Differential Electron Emission

Let us now demonstrate how dielectronic processes influence the differential
electron emission. For this purpose we can study the integrands in (3.209,210)
with the same modifications as required by (3.211,212). Again, ignoring the
different limits of integration, we obtain:

$$\frac{d^2\sigma_D/d\epsilon d\theta}{d^2\sigma_{Sc}/d\epsilon d\theta} \approx \frac{N_P\left\{1 - |F_{o,o}|^2\right\}}{|Z_P - N_P F_{o,o}|^2}. \tag{6.19}$$

Since the maximum electron energies for dielectronic and screened nuclear
processes are large, e.g., T/M_p and $4T/M_p$, respectively, for purposes of
demonstration we will again assume that ϵ ranges from 0 to ∞ which corresponds to $F_{oo} \to 1$ and 0, respectively. Thus (6.19) implies that the dielectronic to screened nuclear cross section ratio decreases from ∞ to $1/Z_p$ for
fast neutral atom impact, whereas it increases from 0 to $(Z_p - q)/Z_p^2$ for fast
ion impact with charge state q.

Thus, based on a simplistic picture, for target ionization it is predicted
that dielectronic processes should be more important than $e - n$ processes at
all electron emission energies for hydrogen atom impact, $Z_p = 1$. For heavier

atom impact, they should dominate the low-energy emission but are less important than $e-n$ processes for higher-energy emission. For fast ion impact, dielectronic processes should never dominate but will achieve a maximum importance at intermediate to low energy electron emission energies.

For projectile ionization, dielectronic processes also produce an enhancement in the low-energy projectile frame electron emission. In the laboratory frame this results in an enhanced electron loss peak intensity. Secondly, just as $e - n$ processes leading to projectile ionization can be viewed as elastic scattering of a projectile electron from the target, [*Burch* et al. (1973), *Wilson* and *Toburen,* (1973)] dielectronic processes leading to the emission of a projectile electron can be viewed as inelastic scattering from the target. The major difference between these two processes is that the inelastic scattering process affects only the target electron, whereas the elastic scattering is from the target as a whole. Hence, in the laboratory frame, $e-n$ processes produce an electron-loss peak centered at $v_e \approx v$, whereas dielectronic processes produce electrons having lower energies. This may result in asymmetric electron loss peaks and may shift the maximum intensity to a slightly lower laboratory energy than expected from a simple elastic scattering model. Note that these asymmetries and shifts are different than those discussed in Sect. 6.4.

In the very first experimental study of differential cross sections for light dressed ion impact, *Wilson* and *Toburen* (1973) observed an asymmetry on the low-energy side of the electron-loss peak at specific laboratory emission angles. They attributed this asymmetry to the presence of dielectronic interactions and attempted to extract the monoelectronic and the dielectronic contributions to the projectile ionization by fitting Gaussian line shapes to the peak, as shown in Fig. 6.21. As no further attempts have been made to experimentally separate the electron-loss peak into its two components, their conclusions have never been confirmed or tested.

Some years later in a study of 15–150 keV H + He collisions, *Rudd* et al. (1980) noted that the electron-loss peak was shifted toward lower energies. However, no analysis was performed as their interest was in lower emission energies. With decreasing emission energy they found that the cross sections increased for dressed ion impact, an unexpected result since this behavior was not observed for bare ion impact. They attributed the increasing cross section as an indication that dielectronic processes are active in the dressed ion collisions. This constitutes the first example of dielectronic processes contributing to ionization of the target.

During the 1980s several attempts were made to theoretically describe the differential electron emission resulting from dielectronic interactions. *Jakubaßa* (1979) used the electron impact approximation, EIA, to study projectile ionization for H impact on heavy targets. *Manson* and *Toburen* (1981) applied the first Born approximation, B1, to target and projectile ionization in He^+ + He collisions. Somewhat later, *Hartley* and *Walters* (1987) used an impulse approximation, IA, combined with the first Born approximation to study the screened nuclear and dielectronic contributions to projectile ioniza-

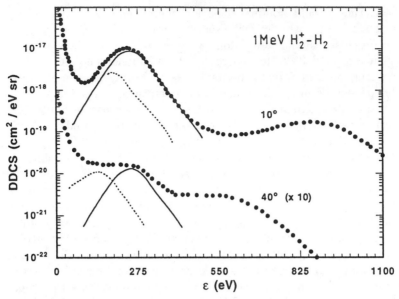

Fig. 6.21. Doubly-differential cross sections for electron emission in 1-MeV H_2^+ + H_2 collisions. Data are those of *Wilson* and *Toburen* (1973). The *dashed* and *solid* *curves* are Gaussian fits to the electron-loss peak which were attributed to e-e and e-n interactions leading to projectile ionization

tion in collisions with heavy targets. As no detailed experimental information relating to dielectronic processes was available, comparisons between experiment and theory tended to concentrate on electron-loss peak intensities, positions, and widths.

Zouros et al. (1995) used the impulse approximation to calculate the differential electron emission which results via dielectronic interactions to target ionization and projectile excitation in 1.5 MeV/u C^{5+} + H_2 collisions. Target electron emission cross sections at 0° and 180° were calculated for projectile excitation to various levels. Their predicted results implied that $e - e$ processes leading to target ionization are considerably weaker than $e - n$ ionization processes. This is easily explained if one recalls that the target $e - n$ ionization process scales as Z_p^2, (25 in this example), whereas the $e - e$ target ionization processes are proportional to the number of projectile electrons. Their work indicates that further studies of this type, particularly for lighter systems, could provide interesting and unique data.

6.5.4 Coincidence Studies of Electron Spectra

To achieve more detailed information about dielectronic processes, *DuBois* (1985) performed coincidence experiments which isolated and identified the target and projectile electron emission components. His first study was limited to 20° electron emission for 0.4 MeV/u He^+ impact on argon. Additional measurements were then made for higher energies and other collision systems [*DuBois* and *Manson* (1986, 1990)]. Typically these studies were performed for forward emission angles, 20° and 30°. Similar studies, performed later at the University of Frankfurt, investigated the entire angular range and studied H and He atom impact on He, Ar, and Kr targets [*Heil* et al. (1991a,b, 1992), *Kuzel* (1991), *Trabold* et al. (1992)]. More recently, *DuBois* (1994) extended these studies to heavier projectiles, i.e., to 0.15 MeV/u carbon particles impacting on helium. Information pertaining to electron emission associated with single ionization of the projectile was obtained, and for neutral carbon impact, the double and triple projectile ionization channels were studied.

Before presenting examples of these data, it is important to recall that the emitted electron-ionized projectile coincidence signals can result from three different ionization channels: 1) projectile ionization induced by the screened target nucleus, $e_P - n_T$, which results in a broad electron-loss peak centered at $v_e \approx v$ in the laboratory frame; 2) electron emission from both the target and the projectile resulting from dielectronic interactions, $e_P - e_T$, which produces electrons having velocities between approximately 0 and v; and 3) from independent second-order screened nuclear interactions, $e_P - n_T, e_T - n_P$, which lead to electrons being emitted with velocities between 0 and $2v$. Thus, by careful investigation of the coincidence electron emission spectrum, it is sometimes possible to determine which process is active.

Experimental and theoretical DDCS for $He^+ + He$ collisions are shown in Fig. 6.22. The experimental cross sections are for the sum of all ionization channels (obtained using non-coincidence techniques) and for the sum of the channels discussed above (obtained from electron-ionized projectile coincidences). Note that the coincidence cross sections do not rapidly decrease toward zero for emission velocities much larger than v. This is direct evidence that higher-order independent ionization processes are active. As seen for the 0.5 MeV/u $He^+ + He$ system, the contribution to the binary-electron emission attributable to higher-order processes is approximately 15%.

Schiwietz (1994) found that first- and second-order $e - n$ target ionization cross sections are comparable in magnitude at all electron emission energies. Thus, theoretical cross sections for the second-order $e - n$ process that are plotted in part (a) of the figure were obtained by multiplying the first-order theoretical $e-n$ cross sections of *DuBois* and *Manson* (1986, 1990) by 0.15. B1 predictions for the other ionization channels are also shown. As seen, inclusion of the second-order $e - n$ process greatly improves the agreement between experiment and theory for both low- and high-energy electron emission.

Additional information about the relative importance of mutual ionization processes with respect to the differential electron emission is shown in Fig. 6.23. Data are for 0.5 MeV H impact on argon *Kuzel*, 1991). Note that in these interactions, mutual ionization of both collision partners accounts for all of the low-energy electron emission in the backward direction, and most of it in the forward direction. Also note that second-order $e - n$ account for approximately 33% of the binary-encounter emission intensity at 30°.

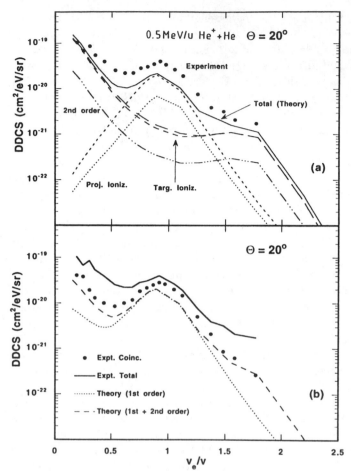

Fig. 6.22a,b. Experimental and theoretical DDCS for electron emission in 0.5 MeV/u He$^+$ + He collision. (a) shows the cross sections for the sum of all processes leading to electron emission,•, exp.: *solid line*, B1 theory, *DuBois* and *Manson* (1990). Also first- and second-order ionization processes are shown. The first-order processes are B1 calculations where the target (or projectile) is ionized and the collision partner remains in the ground state (*short dashed* and *dotted curves* respectively) or is excited (*long* and *medium dashed curves* respectively). From *DuBois* and *Manson* (1990). The second-order calculation for ionization of each collision partner by the screened nuclear charge of the other is from *Schiwietz* (1994). In (b) the experimental cross sections again show the total differential electron emission, *solid curve*, and cross sections for processes where both the target and the projectile are ionized,• . The theoretical curves are for the sum of all processes leading to mutual ionization of both collision partners where only first-order dielectronic processes are considered (*dotted curve*) and where second-order projectile electron-screened target nucleus-target electron-screened projectile nucleus processes (*long dashed curve*) are also considered

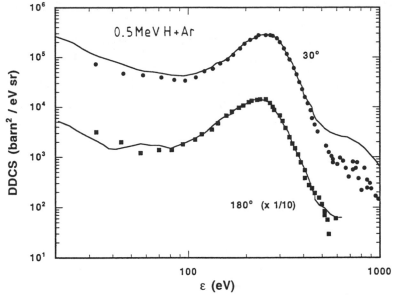

Fig. 6.23. Experimental DDCS for forward and backward electron emission in 0.5 MeV H + Ar collisions. The solid curves are for all processes leading to electron emission while the data points are for the sum of first-order dielectronic processes, the projectile electron-screened target nucleus monoelectronic process, and the second-order projectile electron-screened target nucleustarget electron-screened projectile nucleus process. Data from *Heil* et al. (1991b)

Different theoretical approaches have been applied to calculating projectile ionization [*Hartley* and *Walters* (1987), *Jakubaßa-Amundsen* (1992), *Wang* et al. (1992), *Kuzel* et al. (1993a, 1994)]. *Jakubaßa-Amundsen* (1992) compared several of these methods for H, He and He$^+$ impact on He. For forward emission, the second Born approximation provided the best agreement with experimental data whereas the impulse approximation yielded the poorest results. However, for emission in the backward direction where dielectronic processes are especially important in this highly asymmetric collision system, the Born treatment underestimated the magnitude of these processes. At backward angles, the electron impact approximation yielded the best agreement. A possible reason for these seemingly inconsistent results is that the various theoretical models do not agree on the importance of dielectronic processes, as shown in Fig. 6.24.

In the backward direction, emission from the projectile was also calculated by *Wang* et al. (1992) using the Born approximation. They calculated electron emission resulting from first-order dielectronic processes and also for a second-order process different from that previously described. The second-order double scattering process that they modeled involved the projectile electron successively colliding with a target electron and a screened target nucleus. They found that including this process produced a significant in-

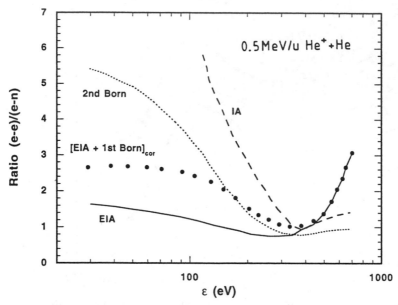

Fig. 6.24. Theoretically predicted fraction of dielectronic processes contributing to projectile ionization in 0.5 MeV/u He$^+$ + He collisions. The theoretical models are the electron impact approximation, EIA, the second-order Born approximation, a coherent addition between the EIA and the first Born approximations, [EIA + B1]cor, and the impulse approximation, IA. From *Jakubaßa-Amundsen* (1992)

crease in the low-energy emission. This again emphasizes the need to include higher-order processes in a proper theoretical treatment.

Equation (6.19) implies that the influence of dielectronic processes is largest for low-energy emission, especially in collisions between energetic neutral atoms. Therefore, *DuBois* and *Manson* (1993, 1994) investigated the differential electron emission for simple dressed projectile-atom collision systems and concentrated on relative changes in the low-energy emission. Cross sections for H, He and He$^+$ impact are compared with those for proton impact in Fig. 6.25.

Note that if only screened nuclear interactions were important, in the region below the electron-loss peak the cross sections should monotonically decrease to values given by q^2 times the proton cross section, q being the net charge of the dressed projectiles, e.g., 0 for H and He impact and 1 for He$^+$ impact. These screening effects are apparent in the region between 10 and 50 eV where the H and He cross sections are smaller than the proton cross section and where the He$^+$ and proton cross sections merge. However, as seen for lower emission energies, the dressed ion impact cross sections monotonically increase. This is attributed to dielectronic processes becoming important and is supported by the observation the increase was observed to systematically disappear when the impact energy was reduced to the threshold energy for the dielectronic process.

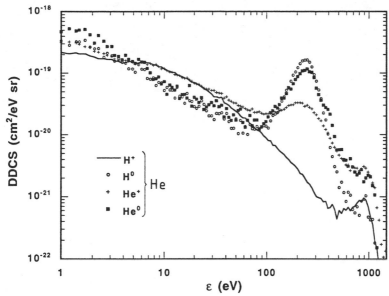

Fig. 6.25. DDCS for electron emission in 500 keV/u H, He, He$^+$, and H$^+$ + He collisions. Experimental data are from *DuBois* and *Manson* (1994)

To summarize, dielectronic processes play a significant role in fast ionizing collisions involving dressed projectiles. Their relative importance is greatest for large impact parameter interactions, hence they primarily lead to low-energy electron emission in the target and the projectile reference frames. For fast atom impact, these processes can dominate; for ion impact, their importance rapidly decreases as the projectile charge state increases. Cross sections for dielectronic processes are expected to scale linearly with the number of bound electrons attached to the ionizing particle. The shortcoming of most theoretical treatments of collisions between dressed particles is their ability to adequately describe these dielectronic processes. Evidence indicates that higher-order independent interactions leading to mutual ionization are also important and that these processes should also be incorporated into theoretical treatments.

7. Multiple Electron Processes

Thus far in this review, most of the processes discussed involve single-electron transitions. Exceptions to this are the two-center dielectronic and the second-order $e - n$ processes discussed in Sect. 6.5. In this final Section, we treat additional multi-electron transitions that can contribute to the electron emission. In contrast to fast proton impact on light targets where single-electron transitions dominate, multiple-electron processes become increasingly more important as Z_p, Z_t, or q increase, or as v decreases.

Although the interest of this review is in fast collisions, meaning those where the velocity of the incoming heavy ion is significantly greater than that of the loosely bound target electrons, a few aspects of slow collisions between heavy ions shall be discussed here. Also, in exceptional cases, Auger electron emission shall be considered. For heavy ion-atom interactions, various multiple-electron channels occur. These include dielectronic transitions for dressed ion impact, multiple outer-shell ionization, inner-shell ionization followed by relaxation processes or accompanied by outer-shell ionization, and transfer ionization processes.

Both experimental and theoretical studies of multiple-electron processes require elaborate techniques. By measuring projectile ion-target ion coincidences, the initial and final charge states of the projectile and the target are established. Moreover, the number of electrons produced via different single- and multiple-electron transition channels can be calculated and the relative importance of multiple-electron transitions determined.

The work concerning multiple-electron processes is so extensive that no attempt can be made to give a complete summary. Rather, we shall select a few typical examples. The reader who is interested in more details is referred to the original articles. An overview of the field of multiple-electron processes can be obtained from the work by *McGuire* and *Weaver* (1977), *Cocke* (1992), *McGuire* (1992), and *Olson* et al. (1993). In particular, developments concerning dynamic electron correlation associated with dielectronic processes have been presented by *McGuire* (1988), *Reading* and *Ford* (1988), and *Stolterfoht* (1991).

In the next section, slow collisions will be discussed. Then, we again treat higher projectile energies which are the primary concern of this work.

7.1 Slow Collisions between Heavy Particles

The mechanisms which dominate at low impact energies are different from those discussed thus far. When the collision velocity becomes significantly smaller than that of the outermost bound electrons, the electron orbitals are modified such that a quasi-molecule is temporarily formed as the collision partners approach one another. This means that the electron emission is influenced by the combined fields of both particles, i.e., by both centers, but in a different way than we have described at high impact energies.

At low impact energies, two mechanisms are primarily responsible for electron emission. The first one, which leads to a continuum of electron energies, occurs via direct coupling between promoted molecular orbitals and the continuum. In this case the theoretical work is rather limited. *Rudd* (1979) and *Woerlee* et al. (1981) developed models for the coupling of molecular orbitals to the continuum yielding exponential expressions for the cross section dependence on the electron energy. An alternative approach to calculate ionization in slow collisions has been provided by *Macek* and *Ovchinnikov* (1994) using the technique of a complex internuclear distance.

The second mechanism promotes electrons from filled inner shells of the quasi-molecule to unfilled outer shells [*Fano* and *Lichten* (1965)]. The subsequent filling of the inner-shell holes is accompanied by the emission of Auger electrons. The promotion mechanisms, which occur in close collisions, are accompanied by the emission of outer-shell electrons. These violent collisions produce many different states which give rise to a broad band of Auger emission lines superimposed on a continuous background (Cacak and Jorgensen, 1970).

An example for the electron emission in heavy-ion atom collisions is given in Fig. 7.1 which shows differential electron emission for 2.5 keV/u argon ion impact on argon [*Rudd* et al. (1966b)]. The structures seen between 5 and 20 eV result from autoionizing transitions whereas those in the vicinity of 200 eV are associated with L-shell Auger emission. *Rudd* et al. (1966b) analyzed the energies of the emission lines and concluded that certain lines, observed at Doppler shifted energies, are associated with emission from the projectile. Unshifted lines are associated with emission from the target. Multi-vacancy states where inner- and outer-shell vacancies are present prior to the Auger transitions, influence the energies of the target and projectile lines. The observation of multiple ionization is consistent with early coincidence studies performed by *Everhart* and *Kessel* (1965, 1966), *Afrosimov* et al. (1964), and *Thomson* (1977) who demonstrated that in slow Ar^{q+}–Ar collisions, the projectile loses several electrons. These studies provided the basis for the conclusion that multiple ionization plays an important role in slow collisions.

Further studies with heavier particles have been performed using the collision system Kr^{q+} + Kr. In this case, a broad structure between 900 and 1200 eV results from L-shell Auger emission, while M-shell Auger electrons occur near 100 eV. The Kr^{q+}–Kr collision system provides a rather extreme case.

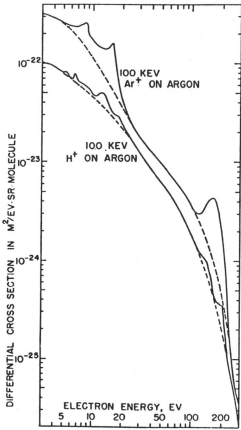

Fig. 7.1. Doubly differential cross sections for 160° electron emission in 100 keV H^+ + Ar and Ar^+ + Ar collisions. Data from *Rudd* et al. (1966)

From coincidence measurements between recoil ions and scattered projectile ions, it was found that $10 - 12$ electrons can be lost from the projectile in a single collision [*Antar* and *Kessel* (1984)]. No information about the number of electrons lost from the target was recorded.

In follow-up studies by *deGroot* et al. (1987, 1988) and *Clapis* and *Kessel* (1990) experiments were performed where Auger electrons were measured in coincidence with the scattered projectile ion. Although in these studies the ion charge state was not measured, the degree of ionization in the collision could be deduced from shifts in the energy of the emission lines from their expected locations.

For example, as shown in Fig. 7.2, the $L_{2,3}$–MM Auger peaks are seen to occur between 900 and 1200 eV rather than between approximately 1200 and 1450 eV as expected for a transition arising from a single vacancy in the L-shell. These shifts result from additional vacancies in the N and M shells

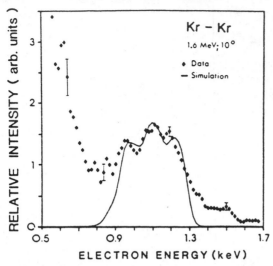

Fig. 7.2. Spectrum of L-MM Auger for 1.6 MeV Kr^{q+}+ Kr collisions. Data points are for electrons measured in coincidence with projectiles scattered at 10° while the *solid line* is a simulated Auger spectrum with the assumption that 13 outer-shell electrons are missing prior to the Auger decay. From *deGroot* et al. (1988)

which are present prior to the Auger transitions. Since each N- and M-shell vacancy reduces the transition energy by roughly 20 and 35 eV, respectively, these data imply that approximately 10−12 electrons are removed in collisions producing a Kr–L vacancy.

To obtain further information about the continuum electron emission, we consider angular distributions of the ejected electrons. Figure 7.3 shows electron spectra for collisions of 50-, 100-, and 400-keV O$^+$ on Ar. The spectra on the left hand side are associated with forward emission angles from 20°–90° while the right figures are due to backward emission angles from 90°–160°. It is important to realize that the projectile energies are so low that the electron-loss and binary-encounter maxima cannot be observed as distinct structures. For all examples, the binary-encounter peak has an energy of less than ≈40 eV. Thus, most of the data shown in Fig. 7.3 are measured above the binary encounter maximum. Unfortunately, not much information is available about the mechanisms producing electrons of such relatively high energies.

Several features can be seen in the electron spectra. The continuum electrons follow approximately an exponential energy dependence predicted by *Rudd* (1979). The most prominent structure near 170 eV can be attributed to the Ar–L Auger electron ejected after the Ar–L vacancy production. The mechanism for producing Ar–L vacancies is due to electron promotion along the molecular orbital 3dσ transiently formed during the collision. This can be verified from molecular-orbital diagrams calculated by *Stolterfoht* (1987b). We also note that normal Ar–L Auger lines attributed to a filled outer shell

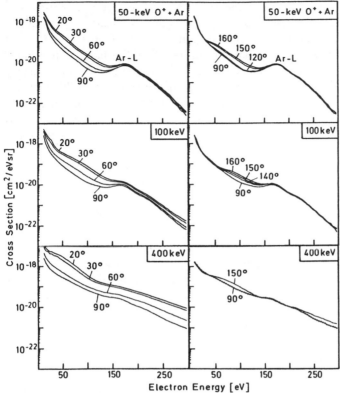

Fig. 7.3. Double-differential cross sections for electron production in 50, 100, and 400 keV O^+ + Ar collisions. The angle given at the curves refers to the electron observation. From *Stolterfoht* and *Schneider* (1979)

are expected near 205 eV. The observed maximum energy of ~170 eV indicates that 2–3 electrons are missing from the outer shell of Ar.

The spectra also show that the Ar–L Auger emission is essentially isotropic. This finding suggests that the life time of the Ar–L vacancy is sufficiently long so that the collision partners have separated before the Auger transition take place, i.e., the Auger electrons are ejected under conditions of spherical symmetry. As noted earlier in this work, Auger electron emission from an isolated atom may, in principle be anisotropic. However, when many states are produced in the collision, experience shows that atomic alignment effects become weak because of statistical averaging effects [*Stolterfoht* (1987a)].

In contrast to the Auger electrons, Fig. 7.3 shows that electron ejection between about 30 to 150 eV is strongly anisotropic. A minimum occurs at 90° whereas the electron intensity is found to be significantly enhanced at forward observation angles. This observation is in accordance with the two-center effects discussed in detail in this work.

The surprising new feature is the increase of the electron emission at backward angles (see the figures at the right side). For instance, at 100 eV the electron intensity observed at 160° is found to be a factor of ~2 higher than the data obtained at 90°. Also it should be noted that close inspection of the 30 − 150 eV region exhibits internal structures which are fixed for varying observation angle, but are shifting with the projectile energy, *Stolterfoht* and *Schneider* (1979). This finding would be inconsistent with a supposition that the 30 − 150 eV region is composed of Auger lines which are influenced by Doppler effects. Hence, at present an obvious explanation for the enhanced backward emission is missing.

It is likely that the emission of the 30 − 150 eV electrons is produced by electron promotion in the quasi-molecule formed during the collision. It should be taken into consideration that electrons which are strongly promoted during the collision occupy σ orbitals [*Fano* and *Lichten* (1965)]. The σ wave functions are "stretched" along the internuclear axis.

Referring back to Fig. 5.25, one notes that the combined Coulomb potential of the collision partners exhibits a symmetry along the internuclear direction which coincides with the beam direction at the end of the collision. If electrons in σ orbitals are removed in the outgoing part of the collision, we would expect an enhanced emission at forward and backward angles, i.e., along the arrow shown in Fig. 5.25. This emission corresponds to a movement in the directions of the internuclear axis where the electrons emerge symmetrically away from the saddle point.

However, to experience the internuclear symmetry, the electron has to oscillate at least a few times around the projectile and target nucleus. This is only possible if the bound electron has a velocity larger than the projectile velocity. The present projectile velocities correspond to electron energies of less than ~30 eV. Assuming that the electron velocity does not change significantly during ionization, it appears plausible that the forward and backward enhancement is seen for electrons with energies $\gtrsim 30$ eV. Alternatively, the enhancement is found to be negligible for electron energies $\lesssim 30$ eV (Fig. 7.3).

It should be kept in mind that the present interpretation involves some speculations. However, the discussion shows that different mechanisms are relevant when slow collisions between heavy ions are studied. Unfortunately, the phenomena of electron emission at low impact energies are not as well understood as those studied for higher impact energies.

7.2 Fractions of Multi-Electron Processes

In this section we return higher projectile energies more relevant for this work. Fist, we consider general properties of multi-electron processes followed by the treatment of individual mechanisms. In Fig. 7.4 characteristic examples of the fraction of ionization of Ne that results from multiple-electron transitions are shown. The data refer to He^+, and He^{2+} impact with energies between 50 and

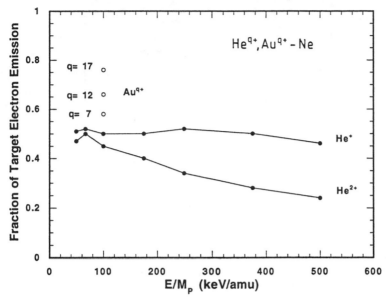

Fig. 7.4. The fraction of multiple ionization relative to the total target electron emission in heavy ion-neon collisions. The He^+, He^{2+} impact data are from *DuBois* (1987, 1989). The Au^{q+} impact data are from *Damsgaard* et al. (1983)

500 keV/u [*DuBois* et al. (1989)] and for 100 keV/u Au^{q+} ions [*Damsgaard* et al. (1983)]. As can be seen even for lighter projectiles such as helium, it is by far not safe to assume that multi-electron processes can be ignored. We also note that multiple outer-shell ionization by light projectiles may be subject to two-center effects in the same way that single outer-shell ionization is.

As an example of the importance of the transfer ionization channel, e.g., where electron capture by the projectile is accompanied by additional target electrons being ionized, see Fig.7.5 Again data are shown for helium *DuBois* (1989) and gold [*Damsgaard* et al. (1983)] ions impacting on a neon target. Note that at low impact energies, virtually all of the target electron emission occurs via the transfer ionization channel.

As the previous two figures demonstrated, total cross section measurements can provide information about the contributions of various multi-electron transition channels. However, detailed information capable of contributing to our understanding of ionization mechanisms is often lost in the total cross sections. Differential electron emission studies can provide the required details, but such studies are rare because the experimental investigations require coincidence techniques which generate small signal rates.

From a theoretical viewpoint, the major difficulty in calculating multi-electron transitions is that many different ionization channels are to be taken into account. Furthermore, the influence of electron correlation should be considered. To circumvent some of these difficulties, theory has generally

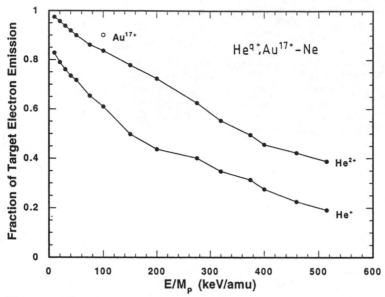

Fig. 7.5. The relative contribution of the transfer ionization channel to the target electron emission in heavy ion-neon collisions. The data are from the same sources as in Fig. 7.4

opted to perform calculations assuming uncorrelated transitions. Deviations between theory and experiment are often interpreted as evidence of electron correlation.

We shall now discuss specific multielectron transition processes and demonstrate what experimental information is currently available. Since relatively little, or no, experimental information exists in many cases, we include studies for proton impact as information obtained from such studies can often be applied to heavy-ion impact.

7.3 Multiple Outer-Shell Ionization

Because ionization of weakly bound electrons generally dominates in ion-atom collisions, we begin with a discussion of multiple outer-shell electron emission. As stated above, multiple electron emission can be a correlated or an uncorrelated process. The uncorrelated process involves two or more independent interactions and, as such, can be described using independent particle models. For example, the uncorrelated process leading to double ionization is when the projectile interacts with two target electrons independently. The double ionization cross section will then be proportional to $P(b)^2$, where $P(b)$ is the probability of ionizing a single electron, whereas the single ionization cross section will be proportional to $P(b)\,[1 - P(b)] \approx P(b)$ for small ionization probabilities.

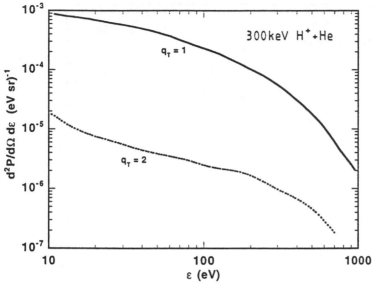

Fig. 7.6. Differential electron emission for single and double target ionization in 300 keV H$^+$ + He collisions. The proton scattering angle was 0.5° and the electron emission angle was 40°. Data are from *Skogvall* and *Schiwietz* (1990)

From Sect. 5.4 we remember that $P(b) \sim (q/v)^2$. Thus, the uncorrelated process becomes negligibly small at high impact velocities. It is important to remember that this statement applies to total cross sections. Portions of the differential cross sections might exhibit uncorrelated behavior at unexpectedly high impact energies, likewise correlated behavior might be exhibited at unexpectedly low impact energies.

Figure 7.6 shows an example of single and double ionization investigated by *Skogvall* and *Schiwietz* (1990) to search for correlated electron emission by investigating the differential probabilities for electron emission by 300 keV proton impact on He. Note that these data are for a fixed projectile scattering angle of 0.5° and electron observation angle of 40°. The relative cross section for double outer-shell ionization is seen to increase with electron emission energy. This general feature will be found regardless of whether correlated or uncorrelated processes are important.

Recently, *Schiwietz* et al. (1994) investigated single and double ionization of helium for 40-MeV proton impact. At this high impact energy, ionization of both target electrons via two independent interactions is expected to be negligibly small. Their measured cross sections and theoretical predictions are shown in Fig. 7.7. We see that the fraction of double ionization again increases with emission energy. Also, as expected, calculated cross sections for double ionization which assumes two independent interactions with the projectile are far smaller than the measured values. However additional calculations assuming a single interaction with the nucleus and which model

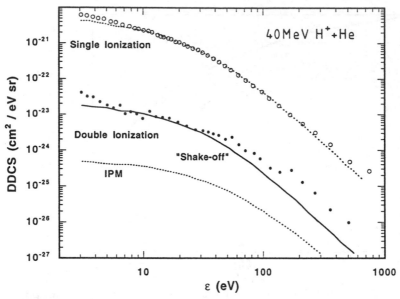

Fig. 7.7. Doubly-differential cross sections for single and double ionization of helium by 40 MeV H⁺ impact. The *dotted curves* assume one or two independent interactions between the proton and the individual target electrons, and the *solid curve* assumes a single interaction leading to double ionization where electron shake off, but no initial state correlation, is included. Data are from *Schiwietz* et al. (1994)

initial state correlation between the two target electrons are found to be in good agreement with the experimental data even above 30 eV.

At lower impact energies, other experimental studies of multiple ionization have also been performed [*Hippler* et al. (1984), *Manzey* (1991), *Chung* and *Rudd* (1996)]. All these studies have attempted to explain their results in terms of uncorrelated emission of electrons, as exemplified in Fig. 7.8. Reasonable agreement is demonstrated between the experimental data of *Hippler* et al. (1984) and predictions based on a simple independent particle model. This agreement was suggested as verification that uncorrelated processes are responsible for the multiple ionization in this example.

Singly-differential cross sections for multiple ionization by 100 keV H⁺ impact on argon and for single- and multiple-electron emission at 90° in 250 keV/u He⁺ + He, Ne collisions were reported by *Chung* and *Rudd* (1996) and *Manzey* (1991), respectively. In both cases, the data demonstrate energy dependencies similar to those shown in Fig. 7.8. Independent electron pictures of multiple outer-shell ionization were also used to interpret these data. We would like to point out that for the He⁺ + Ne system the experiment only investigated electron emission in coincidence with target ions; no information about the transfer ionization channel is available. But Figs. 7.4,5 indicate that transfer ionization accounts for most of the total electron emission in this

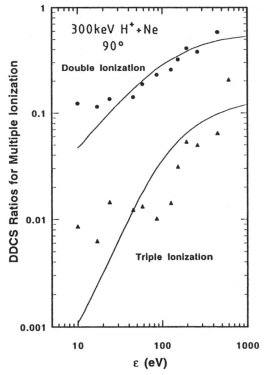

Fig. 7.8. Ratios of multiple to single ionization in 300 keV H^+ + Ne collision. The electron emission angle was 90°. •, ▲ experimental results of *Hippler* et al. (1984); *solid line*: predictions of an independent particle model from *Hippler* et al. (1984)

particular case. Therefore, the interpretation which assumed uncorrelated multiple-outer shell ionization events may be inappropriate.

We are aware of only one example of a theoretical prediction of multiple ionization to the doubly-differential electron emission. *Shinpaugh* et al. (1993) presented nCTMC calculations for 2.4–3.6 MeV/u Xe^{21+} ions impacting on helium and argon targets. Although their general interest was in binary-encounter electron emission, the calculations predict that multiple outer-shell ionization dominates the electron emission for these systems. The theory shows that multiple ionization increases in relative importance as a function of increasing electron emission energy, as discussed.

7.4 Transfer Ionization

As was shown, transfer ionization (TI) can be a major channel for electron emission. This is particularly true for lower impact energies and higher charge state projectiles. In several cases, additional experiments have been performed where the electron emission has been measured in coincidence with

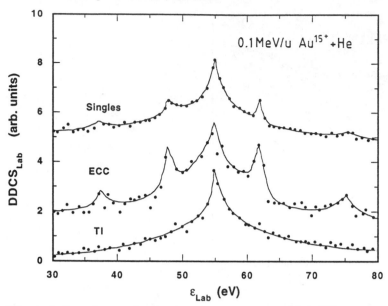

Fig. 7.9. Zero-degree electron emission in 0.1 MeV/u Au^{15+} + He collisions. The upper curve is the singles (non-coincidence) spectrum while the *lower curves* were obtained from coincidences with a selected outgoing projectile charge state. The *solid curves* are fits to the experimental data points. From *Andersen* et al. (1984)

projectiles that captured one or several electrons in the collision. From these data, information about specific ionization mechanisms can be deduced.

For example, *Andersen* et al. (1984) measured $0°$ electron emission in coincidence with the direct ($q_{out} = q_{in}$) and the electron capture channel ($q_{out} = q_{in} - 1$) in 0.1 MeV/u Au^{15+} + He collisions. They observed that the direct channel, e.g., the ECC channel, consisted not only of the usual cusp structure but, in addition, projectile autoionization lines were visible in the wings of the cusp. These autoionization states are populated via a transfer excitation process. The transfer ionization channel, on the other hand, demonstrated a pure cusp-like structure, as shown in Fig. 7.9.

From an analysis of the shape of the TI cusp, *Andersen* et al. (1984) suggested that the transfer ionization was a correlated two-electron transfer to bound projectile states followed by loss of one of the electrons to the continuum by a dielectronic process. In addition, they found that transfer ionization dominated the zero-degree cusp intensity. Later measurements at the Hahn-Meitner Institute Berlin [*Tanis* et al. (1989)] and at the Oak Ridge National Laboratory, *Datz* et al. (1990), supported this correlated electron transfer picture. Impact parameter measurements of *Skutlartz* et al. (1988) and *Plano* et al. (1993) also indicated the importance of correlated electron transfer in heavy ion-atom collisions.

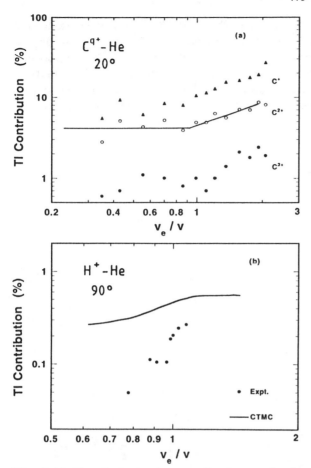

Fig. 7.10. Transfer ionization contributions to the electron emission at 20° for C^{q+} impact (**a**) and for 90° proton impact on helium (**b**). The carbon data are from *DuBois* (1994) while the proton data are from *Hippler* et al. (1987). The solid curve in (**a**) serves to guide the eye whereas in (**b**) it is a theoretical prediction by CTMC calculations based on independent capture and ionization processes

The situation is different for light ion impact. Transfer ionization was found to contribute relatively little to the zero-degree cusp intensity [*Plano* et al. (1993)]. However, it is important to keep in mind that the detection of a process at 0° does not necessarily suggest that it is an important component of the total target ionization. This is because the integration over a solid angle means that only nonzero emission angles contribute to the total cross section. Thus, measurements at emission angles other than 0° are also necessary.

Two TI studies have been performed for nonzero angles of emission. *Hippler* et al. (1987) investigated TI processes leading to 90° electron emission in 300 keV proton-rare gas collisions. *DuBois* (1994) investigated 20° electron emission for the single capture channel in 150 keV/u C^{q+} + He collisions.

Results from these studies are shown in Fig. 7.10. The carbon ion data demonstrate that transfer ionization is a fixed percentage of the differential electron emission for electrons emitted with $v_e < v$. When $v_e > v$ the importance of the TI channel steadily increases. For proton impact, CTMC calculations based on a independent capture and ionization picture indicate this general trend although it is not certain that the experimental data confirm this behavior.

7.5 Inner-Shell Ionization

Finally, we discuss inner-shell ionization as it is often accompanied by multi-electron transition phenomena. For light ion impact, the dominant inner-shell ionization process is the standard Auger process where a single inner-shell electron is ejected and a subsequent Auger process gives rise to the ejection of an outer-shell electron. For heavy ion impact, outer-shell ionization can accompany the initial inner-shell ionization, after which additional outer-shell electrons can be lost in Auger cascade processes.

Experimental information is only available for proton impact ionization of argon. *Sarkadi* et al. (1983) and *Weiter* and *Schuch* (1982) investigated differential electron emission resulting from L- and K-shell ionization, respectively. Relative cross sections for L-shell ionization measured by *Sarkadi* et

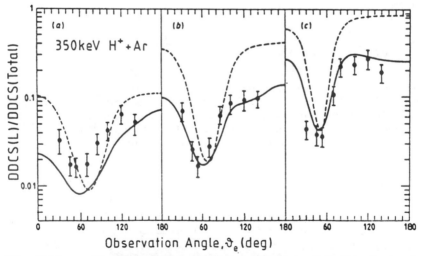

Fig. 7.11a–c. Electron emission associated with argon L-shell ionization by 350-keV H^+ impact. The experimental data refer to cross section ratios for L-shell ionization to total ionization. Theoretical data (*broken curve*) are from *Sarkadi* et al. (1983) who used the Born approximation with screened hydrogenic wave functions. The *solid curves* are theoretical results from *Madison* and *Manson* (1979) who used the Born approximation based on Hartree-Fock-Slater wave functions. From *Sarkadi* et al. (1983)

al. (1983) are shown in Fig. 7.11. General agreement with B1 calculations *Madison* and *Manson* (1979), using Hartree-Fock-Slater wave functions is found. It is seen that the contribution of inner-shell ionization increases with the electron emission energy and for emission in the backward direction. This behavior is partially created by the binary-encounter peak structure in the electron emission from the outer shell. Similar features were also found for K-shell ionization of argon [*Weiter* and *Schuch* (1982)].

Although total cross section studies have demonstrated the important role that multiple ionization plays for heavy ion impact, few experimental or theoretical investigations of the differential electron emission have been performed. The present examples using proton impact show that, whatever the process leading to multiple ionization, the relative importance of multiple ionization increases with increasing electron emission energy.

8. Concluding Remarks

In this work, studies concerning electron production mechanisms in ion-atom collisions, which have been performed over the past 30 years, have been reviewed. This review makes evident the enormous amount of work that has been carried out during that time. Many of the studies have been motivated by the challenge to understand multibody interactions. In addition, many studies have been motivated by the fundamental importance of ion-induced electron emission as applied to many fields, including the technology of ion production, plasma and fusion research, astronomy, and various branches of solid state physics. It is safe to state that the research about ion-induced electron emission has provided a meritorious service to scientists from other fields of physics.

Looking back over the years it is evident that remarkable progress has been made in understanding electron production mechanisms in ion-atom collisions. Various fundamental mechanisms have been identified and successfully modeled. In this process, it was beneficial to consider the concept of collision centers formed when heavy particles strongly interact with electrons. This concept takes into account the specific features of ion-atom collision systems consisting of two heavy particles and one, or more, light particles (i.e., electrons). Using the concept of centers, electron emission in soft and binary collisions can easily be understood as antipodal mechanisms involving a single center associated with the target and the projectile, respectively. Furthermore, various mechanisms can be categorized under the common aspect of being two-center phenomena.

Most characteristic features of ion-atom interactions arise from the specific properties of the Coulomb interaction. Although the Coulomb field involves a long range potential which is formally well known, the theory of ion-atom collisions is by no means easy to develop. The motion of an electron in a two-center field cannot be described without using approximations whose evaluation requires considerable effort. It should also be realized that the long-range force of the Coulomb potential favors situations where the fields of both centers simultaneously play an important role.

Another significant feature of the Coulomb interaction is concerned with the solution of the two-body problem, i.e., the evaluation of the Rutherford formula. The interaction of an electron with a bare nucleus yields the same solutions when described either using classical mechanics or using quantum-mechanical descriptions in first- and in higher-orders. This feature, which may be considered to be an accident, often has important consequences for the field.

First, consequences arise because of the important question of whether it is justifiable to use classical mechanics for the description of ion-atom collisions. It is a fact that classical methods are rather extensively applied in the present field. For example, the binary-encounter theory is used in many applications and the CTMC has gained considerable importance during the past few years. In view of these successes, one may be tempted to believe that binary-encounter electron emission is a classical phenomenon. That this is not true becomes evident when dressed projectiles are used, for then the binary-encounter electrons can exhibit quantum-mechanical diffraction effects. Hence, it should be kept in mind that the applicability of the classical theories primarily stems from the agreement with quantum mechanical theories in the specific case of bare projectiles.

Second, the fact that quantum-mechanical methods yield the same result in first and higher orders has the consequence of extending the applicability of first-order theories. Because of this, surprisingly, the Born approximation is applicable in the case of violent collisions leading to binary-encounter electron emission. In fact, the production of fast electrons by bare projectiles can be described by a simplified version of the Born approximation representing the zero-center case. Hence, this treatment is supported by the fact that first- and higher-order theories yield the same results for bare projectiles.

For dressed ions, however, the projectile center cannot be neglected in violent collisions. In this case, it is favorable to use the binary-encounter theory or the elastic-scattering model. These models allow for a description of the projectile-center case involving the electron-projectile interaction in all orders, but they completely neglect the interaction with the target. In view of this approximation, the elastic-scattering model works surprisingly well for binary-encounter electron emission. At present, the range of its applicability has not yet been fully determined. In this article, an attempt was made to extend the elastic-scattering model and use it for the evaluation of projectile screening effects and electron-loss producing electrons at backward angles.

In recent years the major effort has been on studying two-center effects. The process of electron-capture to the continuum has been investigated by many groups, and detailed information was obtained in order to examine higher-order theories that describe electron interaction involving both centers. Studies of the ECC and saddle-point electron emission seemed to suggest that electrons due to two-center effects occur exclusively in a small range of forward emission angles. Therefore, it was a surprise when experimental studies with fast, highly charged projectiles clearly showed that two-center effects can be observed at all angles. In fact, often two-center effects at backward angles are larger than those at forward angles. Two-center effects were also found in regions of the electron spectra that are traditionally associated with one-center phenomena, i.e., the soft-collision and binary-encounter electrons. Hence, two-center effects can influence the electron emission throughout the entire spectral range.

In view of the large amount of progress which has been achieved in the present field, it is legitimate to speculate what the future will look like. A specific answer to this question cannot be given. In the past few years, surprising new aspects have been discovered for mechanisms that were believed to be fully understood. In light of this, one would anticipate additional new and exciting discoveries. It should also be kept in mind that the investigation of the electron-electron interactions accompanying, or leading to electron emission, is still in its infancy. Moreover, many important branches of the field, such as electron emission in slow ion-atom collisions and the dynamics of electron production in ion-solid interactions, are far from having been fully explored.

Appendix

A. Summary of Analytic Cross Section Formulae

This appendix summarizes analytic formulae providing users with simple tools to estimate ionization cross sections. Also, assistance is provided to convert the atomic units, used in the formulae, into international units. For these purposes we summarize a few expressions which have already been presented within the theoretical framework of the present work.

Total cross sections for ionization may be determined by means of (3.7) given in Sect. 3.1.1:

$$\sigma_{tot} = \frac{Z_p^2\,\pi}{T\,E_b}\left(\ln\frac{2T}{E_b}\right)^{1/2}.\tag{A.1}$$

Recall that Z_p is the projectile charge and E_b is the ionization potential of the electron. The reduced projectile energy is obtained as $T = v^2/2$ where v is the projectile velocity. Expression (A.1) applies for a single electron. If several electrons with the same binding energy are treated, the result has to be multiplied by the number of electrons. If electrons with different binding energies are present, (A.1) has to be applied separately for each group of electrons attributed to a given binding energy, and the results have to be added.

Expression (A.1) is given in atomic units. To apply international units, first convert E_b and T into eV, and then divide by the atomic energy unit H (Hartree), see Table A.1. These values are to be used in (A.1). After the evaluation, the cross sections are converted into international units by multiplying with σ, (Table A.1).

Cross sections differential in electron energy ϵ may be obtained by means of the modified Rutherford formula (3.4) given in Sect. 3.1.1:

$$\frac{d\sigma}{d\epsilon} = \frac{\pi\,Z_p^2}{T\,(cE_b + \epsilon)^2}\tag{A.2}$$

with the empirical constant $c = \ln^{1/2}(2T/E_b)$. This formula also refers to one single electron. For more electrons use the rules described for (A.1). Moreover, the conversion to international units proceed as before: First convert E_b, ϵ, and T into units of eV, and then divide by H to achieve atomic units.

To convert the cross sections results to international units, multiply by σ_ϵ (Table A.1).

Equation (A.2) applies for $\epsilon < 4T$. For values $\epsilon > 4T$ set $d\sigma/d\epsilon = 0$. To obtain a smooth transition near the point $\epsilon = 4T$ one may use the analytic expression (3.80) from Sect. 3.3.4. However, for $\epsilon < 4T$ the results from these formulae are nearly equal (refer back to Fig. 5.8)

Ionization cross sections differential in electron energy ϵ and angle θ may be obtained from (3.79) given in Sect. 3.3.4:

$$\frac{d\sigma}{d\Omega\,d\epsilon} = \frac{Z_p^2}{T}\frac{16}{3\pi\alpha k_c^3}\left[\frac{1}{1+\left(\tilde{K}_m - \tilde{k}\cos\theta\right)^2}\right]^3, \qquad (A.3)$$

where $\tilde{K}_m = (E_b + \epsilon)/\sqrt{4TE_b}$ and $\tilde{k} = \sqrt{\epsilon/E_b}$ are normalized momenta and

$$k_c = \left[2\epsilon + 3E_b\left(\ln\frac{2T}{E_b}\right)^{-2/3}\right]^{1/2}.$$

As before, convert E_b, ϵ, and T into atomic units. To convert the evaluated double-differential cross sections to international units, multiply with σ_ϵ (Table A.1). Expression (A.3) applies for initial 1s states.

For other initial states apply formula (3.91) from Sect. 3.3.4:

$$\frac{d\sigma^{B1-PW}}{d\epsilon d\Omega} = \frac{2Z_p^2}{T\,k_c^3}\,J(k_{iz}), \qquad (A.4)$$

where $k_{iz} = \sqrt{2\epsilon}\cos\theta - (E_b + \epsilon)/\sqrt{2E_b}$. To remove the singularity for $\epsilon \to 0$ in (3.91) we have replaced the momentum k by k_c (given above). Results for the Compton profile $J(k_{iz})$ are available in the literature by *Biggs* et al. (1975).

Triple-differential cross sections for 1s ionization which accounts also for the momentum transfer of the projectile, may be obtained from the B1 formula (3.99) given in Sect. 3.4.2. The expression is lengthy so that it is not repeated here.

When the projectile is dressed, the ionization cross sections are influenced by screening effects. For double differential cross sections one obtain Z_{eff} by means of the elastic scattering model yielding expression (6.11) in Sect. 6.2.3. For single differential cross section one may use expression (6.10) where the fit parameter is set to a mean value $c \approx 3$ relevant for an angle $\lesssim 90°$. For total cross sections it is reasonable to set $\epsilon \approx E_b$, i.e., the ionization energy of the projectile electron. If the projectile carries several electrons, the screening of each electron should be calculated independently and the different contributions are to be summed to obtain Z_{eff}. Note, however, that the proposed methods are still preliminary and they may be improved by means of the elastic scattering model.

Table A.1. Summary of atomic units

Unit	Value
Length (Bohr Radius)	$a_0 = 5.291 \times 10^{-9}$cm
Action (Planck Constant)	$\hbar = 6.521 \times 10^{-16}$eV s
Electron Charge	$e = 1.602 \times 10^{-19}$C
Energy (Hartree)	$H = 27.211$eV
Time	$\tau = 2.418 \times 10^{-17}$s
Velocity	$v_0 = 2.187 \times 10^{8}$cm/s
Total Cross Section	$\sigma = 2.80 \times 10^{-17}$cm^2
Diff. Cross Section	$\sigma_\epsilon = 1.029 \times 10^{-18}$cm^2/eV

In any case it should be kept in mind that the analytic formulae involve various approximations. For instance, they do not account for two-center effects. Note, however, that these effects primarily cause redistributions in the angular space and influence small cross sections. Hence, for many practical cases, one may expect no significant effects on the overall intensity of the ejected electrons.

Also, recall that (A.3,4) originate from an impulsive approximation which is valid for high-electron energies. At low-electron energies, angular distributions are less accurate while the overall intensity is expected to be reliable. If more accurate angular distributions are needed for low-electron energies, one may integrate the B1 formula (3.99) over the momentum transfer. However, numerical integrations require some effort in terms of computer time.

The present formulae may be considered as the simplest approach to total and differential ionization cross sections. Users who need more accurate solutions are referred to the current literature. For instance, reliable values for single-differential cross sections may be obtained from *Rudd* et al. (1992, 1996) who summarized various empirical models. Cross sections which include two-center effects can be evaluated using the CTMC and CDW-EIS codes described in the theoretical sections. To date, these codes can be operated at work stations and personal computers yielding reliable results in a relatively short time.

B. Kinematic Effects on Electron Emission

Kinematic effects for electrons, ejected from a moving emitter, influence electron emission in (i) energy, (ii) angle, and (ii) intensity [*Rudd* and *Macek* (1972), *Stolterfoht* (1987a)]. In the following we give a summary of the formulae which transform the kinematic quantities from the laboratory to the projectile rest frame and vice versa. As usual, primed and unprimed quantities are associated with the projectile and target frame, respectively.

First, let us assume that the double differential cross section $d\sigma(\theta, \epsilon)/d\Omega \, d\epsilon$, with electron emission energy ϵ and angle θ are, known in the laboratory frame of reference. The electron energy in the projectile rest frame is obtained as

$$\epsilon' = \epsilon + T - 2(\epsilon T)^{\frac{1}{2}} \cos\theta, \tag{B.5}$$

where T is the reduced projectile energy, i.e., $T = E_p/M_p$ the projectile energy divided by the projectile-electron mass ratio. The emission angle in the projectile rest frame is

$$\theta' = \arccos\left(\frac{\epsilon - T - \epsilon'}{2\sqrt{T\,\epsilon'}}\right), \tag{B.6}$$

where ϵ' is obtained from (B.5).

Second, let us assume that the double-differential cross section $d\sigma'(\theta', \epsilon')/d\Omega'\, d\epsilon'$ with electron emission energy ϵ' and angle θ' are known in the projectile rest frame. The electron energy in the laboratory rest frame is obtained as

$$\epsilon = \epsilon' + T + 2(\epsilon' T)^{\frac{1}{2}} \cos\theta'. \tag{B.7}$$

The emission angle in the laboratory rest frames is

$$\theta = \arccos\left(\frac{\epsilon + T - \epsilon'}{2\sqrt{T\,\epsilon}}\right), \tag{B.8}$$

where ϵ is obtained from (B.7).

Finally, in the laboratory rest frame, the double-differential cross section is given by:

$$\frac{d\sigma(\theta, \epsilon)}{d\Omega\, d\epsilon} = \left(\frac{\epsilon}{\epsilon'}\right)^{\frac{1}{2}} \frac{d\sigma'(\theta', \epsilon')}{d\Omega'\, d\epsilon'}. \tag{B.9}$$

Similarly, the cross section is obtained in the projectile rest frame.

C. References to Double Differential Cross Section Studies

In this appendix a complete compilation of double-differential cross sections studies for heavy projectile impact is presented in tabular form. The proton impact data are not given here as they have been summarized by *Rudd* et al (1992). The heavy ion data are divided into three groups referring to low (1s to 10s keV/u), intermediate (100s keV/u) and high (MeV/u) projectile energies.

Table C.1. Low impact energies (1s to 10s keV/u)

Collision System	Impact Energy (keV/u)	Electron Energy (eV)	Emission Angle (deg)	References
H^+ / He }-Ar	2.5-10 / 1.25-4	1-26	30-140	*Sataka* et al. (1979a)
H_2^+ / He^+ }-Ar	2.5-10 / 1.25-5	2-26	30, 90	*Sataka* et al. (1979b)
H_2^+ / H_3^+ }-He	2.5-10 / 1.7-6.7	2-50	30-120	*Urakawa* et al. (1981)
H_2^+ / H_3^+ }-He	2.5-15 / 1.7-10	3-200	30-150	*Tokoro* and *Oda* (1985)
He^+ -He	5	5-70	30-150	*Tokoro* et al. (1982)
H / H_2^+ / ^3He / ^4He }-He	15-150 / 7.5-75 / 5-50 / 3.75-37.5	1.5-300	10-160	*Fryar* et al (1977)
He^+ -H_2	7-38	1.5-300	15-160	*Hsu* et al. (1996a)
He^+ -H	7-38	1.5-300	15-160	*Hsu* et al.(1996b)
H -He	15-150	1.5-300	15-160	*Rudd* et al. (1980)
H -H_2O	20-150	1-300	10-160	*Bolorizadeh* and *Rudd* (1986)
He^{2+} -He	60, 100	4-60	17	*Irby* et al. (1988)
He^{2+} -He	50, 100	2-275	0-90	*Bernardi* et al. (1989)

Table C.1. Continued

Collision System	Impact Energy (keV/u)	Electron Energy (eV)	Emission Angle (deg)	References
He^{2+} - $\{$ He / Ne	50, 100 / 100	1–200	0–50	*Bernardi* et al. (1990)
He^{2+} -Ne	106	0.5–440	0–180	*Suárez* et al. (1993a)
He^{2+} - $\{$ He / Ne / Ar	50, 100	3–78	10, 20	*Gay* et al. (1990)
O$^+$ N$^+$ $\}$-Ar	3–31 / 3.6–36	5–500	16–160	*Stolterfoht* and *Schneider* (1979)
Ne$^+$ -Ne	2.5–15	1.5–1000	10–160	*Cacak* and *Jorgensen* (1970)
Ne^{q+} -Ne $q = 1$–4	1.25–40	1.6–1100	45–135	*Woerlee* et al. (1981)
Ar$^+$ -Ar	1.25–7.5	1.5–1000	10–160	*Cacak* and *Jorgensen* (1970)
Ar$^+$ -Ar	2.5	3–250	160	*Rudd* et al. (1966b)
Ar^{2+} -Ar	17.5	100–500	90	*Shanker* et al. (1989)
Kr^{q+} -Kr $q = 2$–5	0.3–8.5	16–1100	45–135	*Gordeev* et al. (1981)
Kr^{q+}-Kr $q = 2$–5	0.6–12	80–1000	45–135	*Gordeev* et al. (1979)
Kr$^+$ Xe $\}$-Kr	1.5–24	100–1500	90	*Clapis* and *Kessel* (1990)

Table C.2. Medium impact energies (100's keV/u)

Collision System	Impact Energy (keV/u)	Electron Energy (eV)	Emission Angle (deg)	References
He –Ar	100	8–80	0	Kuzel et al. (1993a)
H –He	50–500	1–1500	15	DuBois and Manson (1993)
He^{2+} –He	50–250	1–300	15	DuBois (1993)
He$^+$ ⎱ –Ar He^{2+} ⎰	75–500	1–4000	15–125	Toburen and Wilson (1979)
He$^+$ ⎱ –H$_2$O He^{2+} ⎰	75–500	1–4000	15–125	Toburen et al. (1980)
H, He –He, Ar	200	10–400	0	Trabold et al. (1992)
He+He	300	1–3500	15–125	Toburen and Wilson (1977)
He^{2+} –Ne, Ar	300	1–3500	15–125	Toburen and Wilson (1977)
H$_2^+$ –H$_2$	300–750	2–2000	20–125	Wilson and Toburen (1973)
H ⎱ –He He$^+$ ⎰	500	10–500	90–150	Wang at al. (1992)
H$_2^+$ ⎱ –H$_2$, He He$^+$ ⎰	500, 1000 250, 500	20–1000	30–140	Oda and Nishimura (1979)
H, He –He	500	20–1500	30	Heil et al. (1991a)
He$^+$ –He	500	10–1300	20	DuBois and Manson (1986)
H –Ar	500	100–400	10, 170	Jakubaßa (1980)
H –Kr, Xe	500	25–1400	0–180	Kuzel et al. (1994)
H ⎱ –He, Ar He$^+$ ⎰	500	100–450	0–173	Hartley and Walters (1987)
He$^+$ ⎱ –He, Ne Ar ⎰	500	10–700	95–170	Kövér et al. (1988)

Table C.2. Continued

Collision System	Impact Energy (keV/u)	Electron Energy (eV)	Emission Angle (deg)	References
He^+ − { He, Ne, Kr, H_2O, Ar	400–750	10–1500	20, 30 20, 30 20–150	*DuBois* and *Manson* (1990)
H } He } −Ar, Kr	500	20–1100	0–180	*Kuzel* (1991)
H^+, H_2^+, He^+, He^{2+} −N_2	750	20–500	90	*McKnight* and *Rains* (1976)
H_2^+, He^+ −H_2,He	800	50–2000	42	*Kövér* et al. (1980)
H_2^+, He^+ −Ar	800–2000	20–3000	0–60	*Kövér* et al. (1982)
H, He −He	500, 1000	20–1500	0–50	*Heil* et al. (1992)
C^+ − { He, Ne, Ar; CH_4	100	1–800	30, 90	*Toburen* (1979a)
C^+ −He	67–350	10–1500	15–130	*Reinhold* et al. (1990b)
C^+ −He	67–350	2–1500	15–130	*Toburen* et al. (1990)
C^{q+} −He $q=0-3$	100	10–400	30	*DuBois* and *Toburen* (1991)
C^{q+} −He $q=0-4$	100, 150	0.5–500	15, 20	*DuBois* (1994)
H	150	1–500	15	*Toburen* and *DuBois* (1994)
$He^{0,1,2+}$ −He	150	1–500	15	*Toburen* and *DuBois* (1994)
C^{q+} −He $q=0-4$	150	1–500	15	*Toburen* and *DuBois* (1994)

Table C.2. Continued

Collision System	Impact Energy (keV/u)	Electron Energy (eV)	Emission Angle (deg)	References
C^{q+} –Ar, $q = 1\text{–}3$	108–250	1–4000	15–130	Toburen (1979b)
O^{q+} –H_2O, $q = 1\text{–}3$	250	15–1100	15	Toburen (1991)
B^{q+}, C^{q+}, O^{q+}, F^{q+} –He	500	20–1500	10–60	DuBois et al. (1993)
C^{q+} –Ne, Ar, $q = 2\text{–}5$, $Q = 3\text{–}6$	500	20–1500	10–60	DuBois et al. (1993)
I^{6+} –H_2,Ne	200–500	50–1500	17–60	Liao et al. (1992)
$Cu^{3\text{–}5+}$ –H_2	200–500	50–1500	17–60	Liao et al. (1992)
Au^{11+} –He	200–500	50–1500	17–60	Liao et al. (1992)
Au^{q+} –Ar, $q = 9, 11, 23$	200–500	50–1500	17–60	Liao et al. (1992)
Cu^{5+}, Cu^{15+} –He	530	70–2000	0	Jagutzki et al. (1991b)
Au^{11+} –H_2,He	600	50–1800	0–45	Wolff et al. (1993)
I^{7+}, I^{23+}, Au^{11+} –He, Ar	600	50–1800	0–50	Wolff et al. (1992)
Cu^{19+}, I^{23+}, Au^{29+} –Ar	600	100–1800	0–50	Wolff et al. (1993)

Table C.3. High impact energies (MeV/u)

Collision System	Impact Energy (keV/u)	Electron Energy (eV)	Emission Angle (deg)	References
H^+, C^{6+}, N^{7+}, O^{8+}, F^{9+} } $-H_2$,He	1000, 2000	50–1500	0	Lee et al. (1990)
He^{2+}, C^{6+}, O^{8+} } $-$He	1000, 1840	4–2500	20–160	Pedersen et al. (1990, 1991)
$F^{4,6,8,9+}$ $-H_2$	1000	100–2800	0–70	Liao et al. (1994)
O^{q+} $-$Ar, $q = 3$–8	1000	100–1000	180	Breinig et al. (1994)
U^{33+} $-$Ne, Ar	1400	1–4000	20–90	Kelbch et al. (1989a,b)
O^{q+} $-O_2$, $q = 4$–8	1880	5–4000	25–150	Stolterfoht et al. (1974)
I^{23+}, Xe^{21+} } $-$He, Ar	600–3600	100–6000	17–60	Shinpaugh et al. (1993)
O^{q+} $-$Ar, $q = 3$–8	1060–2560	300–2000	90	Burch et al. (1973)
Xe^{21+} $-$He, Ar	2400	100–6000	20	Wang et al. (1993)
$Fe^{17,22+}$ $-$He, Ar	3500	5–8000	27–155	Schneider et al. (1992)

Table C.3. Continued

Collision System	Impact Energy (keV/u)	Electron Energy (eV)	Emission Angle (deg)	References
C^{6+}, O^{8+}, Ne^{10+} } $-$He, Ar	5000	1–5000	20–160	Platten et al. (1987) and Stolterfoht et al. (1995)
Bi^{26+} $-$He, Ar	1400–6000		30–90	Ramm (1994)
Ni^{22+} $-$ { He, Ar	3500		30–90	Ramm (1994)
He, Ar				Ramm (1994)
Pb^{26+} $-$ { Ar, Xe / H$_2$O / SF$_6$ / n-P	6000	100–13000	30	Ramm (1994) / Ramm (1994)
Ne^{7+}, Ar$^{10,18+}$, Kr^{31+}, Xe^{21+}, Au^{24+} } $-$ { He / Ne / Ar / SF$_6$ / n-P	5400–6000		30–140	Ramm (1994)
U^{38+}, Th^{38+} } $-$He, Ar	6000	5–5000	20–150	Schneider et al. (1989)
Mo^{40+} $-$ He	25000	1–5000	20–160	Stolterfoht et al (1987)

n-P = n-Pentan

References

Abrines, R., I.C. Percival (1966a) Proc. L. Phys. Soc. (London) **88**, 861

Abrines R., I.C. Percival (1966b) Proc. R. Phys. Soc. (London) **88**, 873

Afrosimov V.V., Yu.S. Gordeev, M.N. Panov, N.V. Fedorenko (1964) Zh. Tekhn. Fiz. **34**, 1613; Sov. Phys.-Tech. Phys. **9**, 1248 (1965)

Akerib R., S. Borowitz (1961) Phys. Rev. **122**, 1177

Andersen L.H., M. Frost, P. Hvelplund, H. Knudsen, S. Datz (1984) Phys. Rev. Lett. **52**, 518

Anholt R. (1986) Phys. Lett. A **114**, 126

Antar A.A., Q.C. Kessel (1984) Phys Rev A **29**, 1070

Ast H., H.J. Lüdde, R.M. Dreizler (1988) J. Phys. B **21**, 4143

Bachau H., M. Bahri, F. Martín, A. Salin (1991) J. Phys. B **24**, 2015

Bang J., J.M. Hansteen (1959) Mat. Phys. Medd. Dan. Vid. Selesk. **31** (13), p. 43

Banks D., L. Vriens, T.F.M. Bonsen (1969) Proc. R. Phys. Soc. (London) **2**, 976

Bates D.R., G.W. Griffing (1953) Proc. R. Roy. Soc. (London) A **66**, 961

Bates D.R.. G.W. Griffing (1955) Proc. R. Phys. Soc. (London) A **68**, 90

Belkic Dz. (1978) J. Phys. B **11**, 3529

Belkic Dz., R. Gayet, A. Salin (1979) Phys. Rep. **56**, 279

Bell K.L., V. Dose, A.E. Kingston (1969a) J. Phys. B **2**, 129

Bell K.L., V. Dose, A.E. Kingston (1969b) J. Phys. B **2**, 831

Bell F., H. Böckl, M.Z. Wu, H.-D. Betz (1983) J. Phys. B **16**, 187

Berényi D. (1981) Adv. Electron. Electron Phys. **56**, 411

Berényi D., L. Sarkadi, A. Kövér, J. Palinkás (1994) Acta Phys. Hung. **74**, 405

Berkowitz J. (1979) *Photoabsorption, Photoionization, and Photoelectron Spectroscopy* (Academic Press, New York)

Bernardi G., S. Suárez, P. Focke, W. Meckbach (1988), Nucl. Instrum. Methods B **33**, 321

Bernardi G., S. Suárez, P. Fainstein, C. Garibotti, W. Meckbach, P. Focke (1989) Phys. Rev. A **40**, 6863

Bernardi G., P. Fainstein, C.R. Garibotti, S. Suàrez (1990) J. Phys. B **23**, L139

Bernardi G., P. Focke, S. Suàrez, W. Meckbach (1994) Phys. Rev. A **50**, 5338

Bernardi G., W. Meckbach (1995) Phys. Rev. A **51**, 1709

Bernardi G., S. Suárez, D. Fregenal, P. Focke, W. Meckbach (1996) Rev. Sci. Instrum. **67**, 1761

Berry S.D., G.A. Glass, I.A. Sellin, K.O. Groeneveld, D. Hofmann, L.H. Andersen, M. Breinig, S.B. Elston, P. Engar, M.M. Schauer, N. Stolterfoht, H.Schmidt-Böcking, G. Nolte, G. Schiwietz (1985) Phys. Rev. A **31**, 1392

Bethe H. (1930) Ann. Phys. (Leipzig) **5**, 325

Bhalla C.P., R. Shingal (1991) J. Phys. B **24**, 3187

Biggs F., L.B. Mendelsohn, J.B. Mann (1975) At. Data Nucl. Data Tables **16**, 201

Blauth E. (1957) Z. Physik **147**, 1714

Böckl H., F. Bell (1983) Phys. Rev. A **28**, 3207

Boesten L.G.J., T.F.M. Bonsen (1975) Physica B **79**, 292

Boesten L.G.J., T.F.M. Bonsen, D. Banks (1975) J. Phys. B **8**, 628

Bohr N., J. Linhard (1954) K. Dan. Vidensk. Selsk. Mat. Fys. Medd. **28**, 1

Bolorizadeh M.A., M.E. Rudd (1986) Phys. Rev. A **33**, 893

Bonsen T.F.M., L. Vriens (1970) Physica **47**, 307

Bonsen T.F.M., L. Vriens (1971) Physica **54**, 318

Bordenave-Montesquieu A., P. Benoit-Cattin, A. Gleizes, H. Merchez (1976) At. Data Nucl. Data Tables **17**, 157

Brandt D. (1983) Phys. Rev. A **27**, 1314

Brauner M., J.H. Macek (1992) Phys. Rev. A **46**, 2519

Breinig M., S.B. Elston, S. Huldt, L. Liljeby, C.R. Vane, S.D. Berry, G.A. Glass, M. Schauer, I.A. Sellin, G.D. Alton, S. Datz, S. Overbury, L. Laubert, M. Suter (1982) Phys. Rev. A **25**, 3015

Breinig M., J.W. Berryman, F. Segner, D.D. Desai (1994) Phys. Rev. A **50**, 4905

Briggs J.S., K. Taulbjerg (1978) *Structure and Collisions of Ions and Atoms*, in Topics Curr. Phys., Vol. 5, ed. by I.A. Sellin (Springer, Berlin, Heidelberg) p. 105

Briggs J.S., M. Day (1980) J. Phys. B **13**, 4797

Briggs J.S., J.H. Macek (1991) Adv. At. Molec. Opt. Phys. **28**, 1

Brinkman H.C., H.A. Kramers (1930) Proc. Acad. Sci. Amsterdam **33**, 973

Burch D., W.B. Ingalls, J.S. Risley, R. Heffner (1972) Phys. Rev. Lett. **29**, 1719

Burch D., H. Wieman, W.B. Ingalls (1973) Phys. Rev. Lett. **30**, 823

Burgdörfer J. (1983) Phys. Rev. Lett. **51**, 374

Burgdörfer J., M. Breinig, S.B. Elston, I.A. Sellin (1983) Phys. Rev. A **28**, 3277

Burgdörfer J. (1984) in *Forward Electron Ejection in Ion Collisions*, Lecture Notes Phys. Vol. 213, ed. by K.O. Groeneveld, W. Meckbach, I.A. Sellin (Springer, Berlin, Heidelberg) p. 32

Burgdörfer J. (1986) J. Phys. B **19**, 417

Cacak R.K., T. Jorgensen Jr. (1970) Phys Rev. A **4**, 1322

Chen Z., D. Madison, C.D. Lin (1991) J. Phys. B **24**, 3203

Cheshire I.M. (1964) Proc. R. Phys. Soc. (London) **84**, 89

Chew G.F. (1950) Phys. Rev. **80**, 196

Chew G.F., M.L. Goldberger (1952) Phys. Rev. **87**, 778

Chung Y.S., M.E. Rudd (1996) Phys. Rev. A **54**, 4106

Clapis P., Q.C. Kessel (1990) Phys. Rev. A **41**, 4766

Clementi E., C. Roetti (1974) At. Data Nucl. Data Tables **14**, 177

Cocke L. C. (1992) *Electronic and Atomic Collisions, Invited Talks,* ed. by W.R. MacGillivray, I.E. McCarthy, M.C. Standage (Hilger, Bristol) p. 49

Cohen J.S. (1985) J. Phys. B **17**, 1759

Colavecchia F.D., W. Cravero, C.R. Garibotti (1995) Phys. Rev. A **52**, 3737

Coleman J.P. (1968), J. Phys. B **1**, 567

Coleman J.P. (1969) in *Case Studies in Atomic Collision Physics I,* ed. by E.W. McDaniel, M.R.C. McDowell (North Holland, Amsterdam) p. 100.

Compton A.H. (1923) Phys. Rev. **22**, 483

Cravero W., C. Garibotti (1993) Phys. Rev. A **48**, 2012

Cravero W., C. Garibotti (1994) Phys. Rev. A **50**, 3898

Crooks G.B., M.E. Rudd (1970) Phys. Rev. Lett. **25**, 1599

Crothers D.S.F. (1982) J. Phys. B **15**, 2061

Crothers D.S.F., J.F. McCann (1983) J. Phys. B **16**, 3229

Damsgaard H., H.K. Haugen, P. Hvelplund, H. Knudsen (1983) Phys. Rev. A **27**, 112

Datz S., R. Hippler, L.H. Andersen, P.F. Dittner, H. Knudsen, H.F. Krause, P.D. Miller, P.L. Pepmiller, T. Rosseel, R. Schuch, N. Stolterfoht, Y. Yamazaki, C.R. Vane (1990) Phys. Rev. A **41**, 3559

Deco G.R., P.D. Fainstein, R.D. Rivarola (1986) J. Phys. B 11, 213

deGroot P., M.J. Zarcone, Q.C. Kessel (1987) Phys. Rev. A **36**, 2968

deGroot P., M.J. Zarcone, Q.C. Kessel (1988) Phys. Rev. A **38**, 3286

Dettmann K., K.G. Harrison, M.W. Lucas (1974) J. Phys. B 7, 269

Dörner R., V. Mergel, R. Ali, U. Buck, C.L. Cocke, K. Froschauer, O. Jagutzki, S. Lencinas, W.E. Meyerhof, S. Nüttgens, R.E. Olson, H. Schmidt-Böcking, L. Spielberger, K. Tögesi, J. Ullrich, M. Unverzagt, W. Wu (1994) Phys. Rev. Lett. **72**, 3166

Dose V., C. Semini (1974) Helvetica Physica Acta 47, 609.

Drepper F., J.S. Briggs (1976) J. Phys. B 9, 2063

DuBois R.D. (1985) Nucl. Instrum. Methods 10/11, 120

DuBois R.D., S.T. Manson (1986) Phys. Rev. Lett. 57, 1130

DuBois R.D., S.T. Manson, 1987, Phys. Rev. A 35, 2007.

DuBois R.D. (1987) Phys. Rev. A 36, 2585

DuBois R.D., L.H. Toburen (1988) Phys. Rev. A 38, 396

DuBois R.D., À. Kövèr (1989) Phys. Rev. A 40, 3605

DuBois R.D. (1989) Phys. Rev. A 39, 4440

DuBois R.D., S.T. Manson (1990) Phys. Rev. A 42, 1222

DuBois R.D., L.H. Toburen (1991) *XVII Int'l. Conf. on the Physics of Electronic and Atomic Collisions, Abstracts,* ed. by I.E. McCarthy, W.R. MacGillivray, M.C. Standage ISBN 0-7503-0139-2 (Griffith University Press, Brisbane p. 370.

DuBois R.D. (1993) Phys. Rev. A 48, 1123

DuBois R.D., O. Jagutzki, L.H. Toburen, M. Middendorf (1993) Phys. Rev. A 49, 350

DuBois R.D., S.T. Manson (1993) Nucl. Inst. Methods B 79, 93

DuBois R.D. (1994) Phys. Rev. A 50, 364

DuBois R.D., S.T. Manson (1994) Nucl. Instrum. Methods B 86, 161

Duncan M.M., M.G. Menendez, F.L. Eisele (1977) Phys. Rev. A 15, 1785

Eichenauer D., N. Grün, W. Scheid (1981) J. Phys. B 14, 3929

Everhart E., Q.C. Kessel (1965) Phys. Rev. Lett. 14, 247

Everhart E., Q.C. Kessel (1966) Phys. Rev. 146, 27

Fadeev L.D. (1960), Zh. Eksperim. Teor. Fiz. 39, 1459 (*English transl.* 1961: Soviet Phys. – JETP 12, 1014)

Fainstein P.D., R.D. Rivarola (1987) J. Phys. B. 20, 1285

Fainstein P.D., V.H. Ponce, R.D. Rivarola (1987) Phys. Rev. A 36, 3639

Fainstein P.D., V.H. Ponce, R. Rivarola (1988a) J. Phys. B 21, 287

Fainstein P.D., V.H. Ponce, R. Rivarola (1988b) J. Phys. B 21, 2989

Fainstein P.D., V.H. Ponce, R.D. Rivarola (1989a) J. Phys. B 22, 1207

Fainstein P.D., V.H. Ponce, R.D. Rivarola (1989b) J. Physique 50, C1-183

Fainstein P.D., V.H. Ponce, R. Rivarola (1991) J. Phys. B 24, 3091

Fainstein P.D., V.H. Ponce, R.D. Rivarola (1992) Phys. Rev. A 45, 6417

Fainstein P.D., L. Gulyás, A. Salin (1994) J. Phys B 27, L259

Fainstein P.D., L. Gulyás, F. Martin, A. Salin (1996) Phys. Rev A 53, 3243

Fano U. (1961) Phys. Rev. A 124, 1866

Fano U., W. Lichten (1965) Phys. Rev. Lett. 14, 627

Feschbach H. (1962) Ann. Phys. (New York) 19, 287

Fite W.L., R.J. Stebbing, D.G. Hummer, R.T. Brackman (1960) Phys. Rev. **119**, 663

Ford A.L., E.G. Fitchard, J.F. Reading (1977) Phys. Rev. A **16**, 133

Ford A.L., J.F. Reading, R.L. Becker (1982) J. Phys. B **15**, 3257

Fryar J., M.E. Rudd, J.S. Risley (1977) *XI Int'l Conf. on the Physics of Electronic and Atomic Collisions, Abstracts* (Commissariat A L'Energie Atomique, Paris) p. 984

Gaither III C.C., M. Breinig, J.W. Berryman, B.F. Hasson, J.D. Richards, K. Price (1993) Phys. Rev. A **47**, 3878

Gallaher D.F., L. Wilets (1968) Phys. Rev. **169**, 139

Garcia J.D. (1970) Phys. Rev. A **1**, 1402

Garibotti C.R., J.E. Miraglia (1980) Phys. Rev. A **21**, 572

Garibotti C.R., J.E. Miraglia (1981) J. Phys. B **14**, 863

Garibotti C.R., R.O. Barachina (1983) Phys. Rev. A **28**, 2792

Garvey R.H., C.J. Jackman, A.E.S. Green (1975) Phys. Rev. A **12**, 1144

Gay T.J., M.W. Gealy, M.E. Rudd (1990) J. Phys. B **23**, L823

Gilbody H.B., J.V. Ireland (1964) Proc. R. Soc. (London) A**77**, 137

Gillespie G.H., Y.-K. Kim, K.-T. Cheng (1978) Phys. Rev. A **17**, 1284

Gillespie G.H. (1979) Phys. Lett. A **72**, 329

Gillespie G.H., M. Inokuti (1980) Phys. Rev. A **22**, 2430

Gillespie G.H. (1983) Phys. Lett. A **93**, 327

Glauber R.J. (1959) in *Lectures in Theoretical Physics,* Vol. 1, ed. by W.E. Britten, L.G. Duncan (Interscience, New York) p. 315

González A.D., P. Dahl, P. Hvelplund, K. Taulbjerg (1992) J. Phys. B **25**, L573

Gordeev Yu.S., P.H. Woerlee, H. de Waard, F.W. Saris (1979) in *XI Int'l Conf. on the Physics of Electronic and Atomic Collisions, Abstracts* (The Society for Atomic Collision Research, Kyoto, Japan) p. 746

Gordeev Yu.S., P.H. Woerlee, H. de Waard, F.W. Saris (1981) J. Phys. B **14**, 513

Green A.E.S., D.L. Sellin, A. Zachor (1969) Phys. Rev. **184**, 1

Grozdanov T.P., R.K. Janev (1978) Phys. Rev. A **17**, 880

Gryzinski M. (1959) Phys. Rev. **115**, 374

Gryzinski M. (1965) Phys. Rev. **138**, 305, 322, 336

Gulyás L., P.D. Fainstein, A. Salin (1995a) J. Phys B **28**, 245

Gulyás L., P.D. Fainstein, A. Salin (1995b) Nucl. Instrum. Methods B **98**, 338

Güttner F., W. König, B.M. Martin, B. Povh, H. Skapa, J. Soltani, Th. Walcher, F. Bosch and C. Kuzhuharov (1982) Z. Physik A **304**, 207

Hansen J.P., L. Kocbach (1989) J. Phys. B **22**, L71

Hansteen J.H. (1975) Adv. At. and Mol. Phys. **11**, 299

Hansteen, J.M., O.M. Johnson, L. Kocbach (1975) At. Data Nucl. Data Tables **15**, 305

Hardie D.J.W., R.E. Olson (1983) J. Phys. B **16**, 1983

Harrison K.G., M.W. Lucas (1970) Phys. Lett. A **33**, 142

Hartley, H.M., H.R.J. Walters (1987) J. Phys. B **20**, 3811

Heil O. (1991) private communication, Wang et al. (1991)

Heil O., R.D. DuBois, R. Maier, M. Kuzel, K-O. Groeneveld (1991a) Z. Physik D **21**, 235

Heil O., R.D. DuBois, R. Maier, M. Kuzel, K-O. Groeneveld (1991b) Nucl. Instrum. Methods B **56/57**, 282

Heil O., R.D. DuBois, R. Maier, M. Kuzel, K.-O. Groeneveld (1992) Phys. Rev. A **45**, 2850

Henne A., H.J. Lüdde, A. Toepfer, T. Gluth, R. M. Dreizler (1993) J. Phys. B **26**, 3815

Herman F., S. Skillman (1963) *Atomic Structure Calculations* (Prentice-Hall, Englewood Cliffs, NJ)

Hidmi H.I., P. Richard, J.M. Sanders, H. Schöne, J.P. Giese, D.H. Lee, T.J.M. Zouros, S.L. Varghese (1993) Phys. Rev. A **48**, 4421

Hippler R., J. Bossler, H.O. Lutz (1984) J. Phys. B **17**, 2453

Hippler R., G. Schiwietz, J. Bossler (1987) Phys. Rev. A **35**, 485

Horsdal-Pedersen E., L. Larsen (1979) J. Phys. B **24**, 4099

Hsu Y.-Y., M.W. Gealy, G.W. Kerby III, M.E. Rudd (1996a) Phys. Rev. A **53**, 297

Hsu Y.-Y., M.W. Gealy, G.W. Kerby III, M.E. Rudd, D.R. Schultz, C.O. Reinhold (1996b) Phys. Rev. A **53**, 303

Huang K. (1963) *Statistical Mechanics* (Wiley, New York).

Hülskötter H.P., W.E. Meyerhof, E. Dillard, N. Guardala (1990) Phys. Rev. Lett. **63**, 315

Hvelplund P., H. Tawara, K. Komaki, Y. Yamazaki, K. Kuroki, H. Watanabe, K. Kawatsura, M. Sataka, M. Imai, Y. Kanai, T. Kambara, Y. Awaya (1991) J. Phys. Soc. Jpn. **60**, 3675

Inokuti M. (1971) Rev. Mod. Phys. **43**, 297

Irby V.D., T.J. Gay, Wm. Edwards, E.B. Hale, M.L. McKenzie. R.E. Olson (1988) Phys. Rev. A **37**, 3612

Irby V.D., S. Datz, P.F. Dittner, N.L. Jones, H.F. Krause, C.R. Vane (1993) Phys. Rev. A **47**, 2957

Irby V.D. (1995) Phys. Rev. A **51**, 1713

Itoh A., T. Schneider, G. Schiwietz, Z. Roller, H. Platten, G. Nolte, D. Schneider, N. Stolterfoht (1983) J. Phys. B **16**, 3965

Itoh A., D. Schneider, T. Schneider, T.J.M. Zouros, G. Nolte, G. Schiwietz, W. Zeitz, N. Stolterfoht (1985) Phys. Rev. A **31**, 684

Jackson J.D. (1962) *Classical Electrodynamics* (Wiley, New York) p. 451

Jagutzki O., R. Koch, A. Skutlartz, C. Kelbch, H. Schmidt-Böcking (1991a) J. Phys. B **24**, 993

Jagutzki O., S. Hagmann, H. Schmidt-Böcking, R.E. Olson, D.R. Schultz, R. Dörner, R. Koch, A. Skutlartz, A. Gonzáles, T.B. Quinteros, C. Kelbch, R. Richard (1991b) J. Phys. B **24**, 2579

Jakubaßa D.H. (1980) J. Phys. B **13**, 2099

Jakubaßa-Amundsen D.H. (1983) J. Phys. B **16**, 1767

Jakubaßa-Amundsen D.H. (1988) Phys. Rev. A **38**, 70

Jakubaßa-Amundsen, D.H. (1992) Z. Physik D **22**, 701

Janev R.K., L.P. Presnyakov (1980) J. Phys. B **13**, 4233

Kabachnik, N.M. (1993) J. Phys. B **26**, 3803

Kelbch C., R.E. Olson, S. Schmidt, H. Schmidt-Böcking, S. Hagmann (1989a) Phys. Lett. A **139**, 304

Kelbch C., R.E. Olson, S. Schmidt, H. Schmidt-Böcking, S. Hagmann (1989b) J. Phys. B **22**, 2171

Keller N. (1989) Diploma Thesis, Institut für Kernphysik der J.W. Goethe Universität Frankfurt (unpublished).

Kerby III G.W., M.W. Gealy, Y.-Y Hsu, M.E. Rudd, D.R. Schultz, C.O. Reinhold (1995) Phys. Rev. A **51**, 2256

Kim Y.-K. (1983) Phys. Rev. A **28**, 656

Knudsen H., L. H. Andersen, P. Hvelplund, G. Astner, H. Cederquist, H. Danared, L. Liljeby, K-G. Rensfelt (1984) J. Phys. B **17**, 3545

Kocbach L., J.S. Briggs (1984) J. Phys. B. **17**, 3255

Kövér A., S. Ricz, Gy. Szabó, D. Berényi, E. Koltay, J.Végh (1980) Phys. Lett. A **79**, 305

Kövér A., Gy. Szabó, D. Berényi, D. Varga, I. Kádár, J. Végh (1982) Phys. Lett. A **89**, 71

Kövér A., D. Varga, Gy. Szabó, D. Berényi, I. Kádár, S.A. Ricz, J. Végh, G. Hock (1983) J. Phys. B **16**, 1017

Kövér A., Gy. Szabó, D. Berényi, L. Gulyás, I. Cserny, K.-O. Groeneveld, D. Hofmann, P. Koschar, M. Burkhard (1986) J. Phys. B **19**, 1187

Kövér A., Gy. Szabó, L. Gulyás, K. Tökési, D. Berényi, O. Heil, K.-O. Groeneveld (1988) J. Phys. B **21**, 3231

Kövér A., L. Sarkadi, J. Pálinkás, D. Berényi, Gy. Szabó, T. Vajnai, O. Heil, K.-O. Groeneveld, J. Gibbons, I.A. Sellin (1989) J. Phys. B **22**, 1595

Kuyatt C.E., T. Jorgensen Jr. (1963) Phys. Rev. **130**, 1444

Kuzel M. (1991) Diploma Thesis, Institute für Kerphysik der J.W. Goethe Universität, Frankfurt

Kuzel M., L. Sarkadi, J. Pálinkás, P.A. Závodszky, R. Maier, D. Berényi, K.-O. Groeneveld (1993a) Phys. Rev. A **48**, R1745

Kuzel M., R. Maier, O. Heil, D.H. Jakubassa-Amundsen, M.W. Lucas, K-O. Groeneveld (1993b) Phys. Rev. Lett. **18**, 2879

Kuzel M., R.D. DuBois, R. Maier, O. Heil, D.H. Jakubassa-Amundsen, M.W. Lucus, K-O. Groeneveld (1994) J. Phys. B **27**, 1993

Landau D.L., E.M. Lifschitz (1958) *Quantum Mechanics: Non Relativistic Theory* (Addison-Wesley, Reading, MA) p. 459

Lee D.H., T.J.M Zouros, J.M. Sanders, J.L. Shinpaugh, T.N. Tipping, S.L. Varghese, B.D. DePaola, P. Richard (1989) Nucl. Instr. Methods in Phys. Res. B **40/41**, 17

Lee D.H., P. Richard, T.J.M. Zouros, J. M. Sanders, J.L. Shinpaugh, H. Hidmi (1990) Phys. Rev. A **41**, 4816

Liao C., S. Hagmann, C.P. Bhalla, R. Shingal, H. Schmidt-Böcking, R. Mann, J. Shinpaugh, W. Wolff, H. Wolf (1992) in *VIth Int'l Conf. on the Physics of Highly Charged Ions,* ed. by P. Richard, M. Stöckli, C.L. Cocke, C.D. Lin, AIP Conf. Proc. **274**, (American Institute of Physics, New York) p. 281

Liao C., P. Richard, S.R. Grabbe, C. Bhalla, T.J.M. Zouros, S. Hagmann (1994) Phys. Rev. A **50**, 1328

Lucas M.W., D.H. Jakubaßa-Amundsen, M. Kuzel, K.O. Groeneveld (1997) Int. J. Mod. Phys. A Vol. 12, 305

Lüdde H.J., H. Ast, R.M. Dreizler (1987) Phys. Lett. A **125**, 197

Lüdde H.J., H. Ast, R.M. Dreizler (1988) J. Phys. B **21**, 4131

Lüdde H.J., R.M. Dreizler (1989 J. Phys. B 22, 3254

Lüdde H.J., A. Henne, A. Salin, A. Toepfer, R.M. Dreizler (1993) J. Phys. B **26**, 2667

Macek J. (1970) Phys. Rev. A **1**, 235

Macek J., R. Shakeshaft (1980) Phys. Rev. A 22, 1441

Macek J., K. Taulbjerg (1981) Phys. Rev. Lett. **46**, 170

Macek J. (1991) Nucl. Instrum. Methods B **53**, 416

Macek J., K. Taulbjerg (1993) J. Phys. B **26**, 1353

Macek J.H., S.Yu. Ovchinnikov (1994) Phys. Rev. A **50**, 468

Macías A., F. Martín, A. Riera, M. Yáñez (1987) Phys. Rev. A **36**, 4179

Macías A., F. Martín, A. Riera, M. Yáñez (1988) Int. J. Quantum Chem. **33**, 279

Madison D.H. (1973) Phys. Rev. A **8**, 2449

Madison D.H., E. Merzbacher (1975) in *Atomic Inner-Shell Processes,* ed. by B. Crasemann (Academic Press, New York) p. 1

Madison D.H., S.T. Manson (1979) Phys. Rev. A **20**, 825

Madsen J.H., K. Taulbjerg (1994) J. Phys. B 27, 2239

Madsen J.N., K. Taulbjerg (1995) J. Phys. B 28, 1251

Maidagan, J.M., R.D. Rivarola (1984) J. Phys. B 17, 2477

Manson S.T., L.H. Toburen, D.H. Madison, N. Stolterfoht (1975) Phys. Rev. A 12, 60

Manson S.T., L.H. Toburen (1981) Phys. Rev. Lett. 46, 529

Manson S.T., R.D. DuBois (1992) Phys. Rev. A 46, R6773

Manzey D. (1991) Diploma Thesis, J.W. Goethe Universität Frankfurt (unpublished)

Martín F., A. Salin (1995) J. Phys. B 28, 639

Martir M.H., A.L. Ford, J.F. Reading, R.L. Becker (1982) J. Phys. B 15, 1729

Massey H.S.W., C.B.O. Mohr (1933) Proc. Roy Soc. (London) A 140, 613

Massey H.S.W., (1956) in *Handbuch der Physik*, Vol. 36, edited by S. Flügge (Springer Verlag, Berlin) p. 356

McCartney M., D.S.F. Crothers (1993) J. Phys. B 26, 4561

DcDaniel E.W., J.B.A. Mitchel, M.E. Rudd (1993) *Atomic Collisions: Heavy Particle Projectiles* (Wiley, New York)

McDowell M.R.C. (1961) Proc. R. Soc. (London) A 264, 277

McDowell, M.C.R., J.P. Coleman (1970) *Introduction to the Theory of Ion Atom Collisions* (North Holland, Amsterdam)

McGuire J.H., L. Weaver (1977) Phys. Rev. A 16, 41

McGuire, J.H., N. Stolterfoht, P.R. Simony (1981) Phys. Rev. A 24, 97

McGuire J.H. (1982) Phys. Rev. A 26, 143

McGuire J.H. (1983) J. Phys. B 16, 3805

McGuire J.H. (1988) *High-Energy Ion-Atom Collisions*, Lecture Notes Phys. Vol. 294, ed. by D. Berényi, G. Hock (Springer, Berlin, Heidelberg) pp. 415

McGuire J.H. (1992) Adv. Atom. Molec. Opt. Phys. 29, 217

McGuire J.H. (1997) *Introduction to Dynamic Correlation: Multiple Electron Transitions in Atomic Collisions.* (Cambridge University Press, Cambridge) (in press)

McKnight R.H., R.G. Rains (1976) Phys. Lett. A 57, 129

Mehlhorn W. (1985) *Atomic Inner-Shell Processes,* ed. by B. Crasemann (Plenum, New York) p. 119

Meckbach W., B. Nemirovsky, C. Garibotti (1981) Phys. Rev. A 24, 1793

Meckbach W., P.R. Focke, A.R. Goñi, S. Suàrez, J. Macek, M.G. Menendez (1986) Phys. Rev. Lett. 57, 1587

Meckbach W., S. Suàrez, P. Focke, G. Bernardi (1991) J. Phys. B 24, 3763

Menendez M.G., A. Huetz, M.M. Duncan (1991) *Proc. Conf. on High-Energy Ion-Atom Collisions*, Lecture Notes Phys., Vol. 376, ed. by D. Berényi, G. Hock (Springer, Berlin, Heidelberg) p. 23

Messiah A. (1962) *Quantum Mechanics*, Vol. I (North Holland, Amsterdam)

Meyerhof W.E., K. Taulbjerg (1977) Ann. Rev. Nucl. Sci. **27**, 279

Miraglia J.E., R.D. Piacentini, R.D. Rivarola, A. Salin (1980) J. Phys. B **14**, L197

Miraglia J.E. (1983) J. Phys. B **16**, 1029

Miraglia J.E., J. Macek (1991) Phys. Rev. A **43**, 5919

Moe, D., E. Petsch (1958) Phys. Rev. **110**, 1358

Montemayor V.J., G. Schiwietz (1989) J. Phys. B**22**, 2555

Montenegro E.C., W.S. Melo, W.E. Meyerhof, A.G. de Pinho (1992) Phys. Rev. Lett. **69**, 3033

Montenegro E.C., A. Belkacem, D.W. Spooner, W.E. Meyerhof, M.B. Shah (1993a) Phys. Rev. A **47**, 1045

Montenegro E.C., W.S. Melo, W.E. Meyerhof, A.G de Pinho (1993b) Phys. Rev. A **48**, 4259

Montenegro E.C., W.E. Meyerhof, J.H. McGuire (1994) Adv. Atom. Molec. Opt. Phys. **34**, 249

Moshammer R., J. Ullrich, M. Unverzagt, W. Schmidt, P. Jardin, R.E. Olson, R. Mann, R. Dörner, V. Mergel, U. Buck, H. Schmidt-Böcking (1994) Phys. Rev. Lett. **73**, 3371

Moshammer, R., J. Ullrich, H. Kollmus, W. Schmidt, M. Unverzagt, O. Jagutzki, V. Mergel, H. Schmidt-Böcking, R. Mann, C.J. Wood, R.E. Olson (1996) Phys. Rev. Lett. **77**, 1241 1242

Mott N.F., H.S.W. Massey (1949) *The Theory of Atomic Collisions*, 2nd ed., (Oxford University Press, New York) p. 234

Oda N., F. Nishimura (1979) in *XI Int'l Conf. on the Physics of Electronic and Atomic Collisions, Abstracts* (The Society for Atomic Collision Research, Kyoto, Japan) p. 622

Oldham W.J.B. (1967) Phys. Rev. **161**, 1

Olson R.E., A. Salop (1976) Phys. Rev. A **14**, 579

Olson R.E., A. Salop (1977) Phys. Rev A **16**, 531

Olson R.E. (1980) J. Phys. B **13**, 483

Olson R.E. (1983) Phys. Rev. A **27**, 1871

Olson R.E. (1986) Phys. Rev. A **33**, 4397

Olson R.E., T.J. Gay, H.G. Berry, E.B. Hale, V.D. Irby (1987) Phys. Rev. Lett. **59**, 36

Olson R.E., C.O. Reinhold, D.R. Schultz (1990) J. Phys. B **23**, L455

Olson R.E., J. Wang, J. Ullrich (1993) in *Proc. XVIII Int'l Conf. on the Physics of Electronic and Atomic Collisions* ed. by T. Andersen, B. Fastrup, F. Folkman, H. Knudsen, N. Andersen (American Institute of Physics, New York) p. 520

Oppenheimer J.R. (1928) Phys. Rev. **31**, 349

Oswald N., R. Schramm, D.H. Jakubaßa-Amundsen, H.-D. Betz (1994) J. Phys. B **27**, 513

Ovchinnikov S.Yu., D.B. Khrebtukov (1987) in *15th Int'l Conf. on the Physics of Electronic and Atomic Collisions, Abstract of Contributed Papers,* ed. by J. Geddes H.B. Gilbody, A.E. Kingston, C.J. Latimer, H.J.R. Walters (Univ. Press, Brighton, UK) p. 596

Park J.T., J.E. Aldag, J.M. George, J.L. Peacher (1977) Phys. Rev. A **15**, 508

Pedersen J.O., P. Hvelplund, A.G. Petersen, P.D. Fainstein (1990) J. Phys. B **23**, L597

Pedersen J.O., P. Hvelplund, A.G. Petersen, P.D. Fainstein (1991) J. Phys. B **24**, 4001

Pieksma M., S.Y. Ovchinnikov (1994) J. Phys. B **27**, 4573

Pieksma, M., S. Y. Ovchinnikov, J. van Eck, W. B. Westerveld and A. Niehaus, 1994, Phys. Rev. Lett. 73, 46.

Plano V.L., L. Sarkadi, P. Závodszky, D. Berényi, J. Pálinkás, L. Gulyás, E. Takács, L. Tóth, J.A. Tanis (1993) *Proc. VIth Int'l Conf. Physics of Highly Charged Ions,* ed. by P. Richard, M. Stöckli, C.L. Cocke, C.D. Lin, AIP Conf. Proc. **274**, 307

Platten H., G. Schiwietz, T. Schneider, D. Schneider, W. Zeitz, K. Musiol, T.J.M. Zouros, R. Kowallik, N. Stolterfoht (1987) in *XVth Int'l Conf. on the Physics of Electronic and Atomic Collisions, Abstracts of Contributed Papers* ed. by J. Geddes (Univ. Press, Brighton, UK) p. 437

Ponce V.H., P.D. Fainstein, R.D. Rivarola (1993) J. Phys. B **26**, 1343

Pradhan T. (1957) Phys. Rev. **105**, 1250

Pregliasco R.G., C.R. Garibotti, R.O. Barachina (1994) J. Phys. **B 27**, 1151

Prost, M., D. Schneider, R. DuBois, P. Ziem, H.C. Werner, N. Stolterfoht (1982) *Proceedings of the International Seminar on High Energy from Atom Collisions.* Invited Lectures, ed. D. Berényi (Akadémiai Kiado, Budapest) p. 99

Quinteros T.B., A.D. González, O. Jagutzki, A. Skutlartz, D.H. Lee, S. Hagmann, P. Richard, C. Kelbch, S.I. Varghese, H. Schmidt-Böcking, 1991, J. Phys. B **24**, 1377

Ramm U. (1994), Ph.D. Thesis, J.W. Goethe Universität Frankfurt (unpublished)

Ramsauer C. (1921) Ann. Phys. (Leipzig) **66**, 546

Reading J.F., A.L. Ford, E.G. Fitchard (1976) Phys. Rev. Lett. **36**, 573

Reading J.F., A.L. Ford, G.L. Swafford, A. Fitchard (1979) Phys. Rev. A **20**, 130

Reading J.F., A.L. Ford (1979) J. Phys. B **12**, 1367

Reading J.F., A.L. Ford, R.L. Becker (1981) J. Phys. B **14**, 1995

Reading J.F., A.L. Ford (1987) J. Phys. B 20, 3747

Reading J.F., A.L. Ford (1988) in *XVth Int'l Conf. on Electronic and Atomic Collisions, Invited Papers*, ed. by H.B. Gilbody, W.R. Newell, F.H. Read, A.C. Smith (North Holland, Amsterdam) p. 693

Reinhold C.O., C.A. Falcón (1986) Phys. Rev. A 33, 3859

Reinhold C.O. (1987), Ph.D. Thesis (Universidad de Buenos Aires) unpublished

Reinhold C.O., C.A. Falcón (1988a) J. Phys. B 21, 1829

Reinhold C.O., C.A. Falcón (1988b) J. Phys. B 21, 2473

Reinhold C.O., R.E. Olson (1989) Phys. Rev. A 39, 3861

Reinhold C.O., D.R. Schultz (1989) Phys. Rev. A 40, 7373

Reinhold C.O., D.R. Schultz, R. Olson (1990a) J. Phys. B 23, L591

Reinhold C.O. D.R. Schultz, R.E. Olson, L.H. Toburen, R.D. DuBois (1990b) J. Phys. B 23, L297

Reinhold C.O., D.R. Schultz, R.E. Olson, C. Kelbch, R. Koch, H. Schmidt-Böcking (1991) Phys. Rev. Lett. 66, 1842

Reinhold C.O., J. Burgdörfer (1993) J. Phys. B 26, 3101

Richard P., D.H. Lee, T.J.M. Zouros, J.M. Sanders, J.L. Shinpaugh (1990) J. Phys. B 23, L213

Richard P. (1996) in *Two-Center Effects in Ion-Atom Collisions*, ed. by T.J. Gay, A.F. Starace, AIP Conf. Proc. 362, 69

Ricz S., B. Sulik, N. Stolterfoht, I. Kádár (1993) Phys. Rev. A 47, 1930

Rivarola R.D., P.D. Fainstein (1987) Nucl. Instrum. Methods Phys. Res. B 24/25, 240

Rivarola R.D., P.D. Fainstein, V.H. Ponce (1989) in *Proc. XVI Int'l Conf. on the Physics of Electronic and Atomic Collisions*, ed. by A. Dalgarno, R.S. Freund, E.M. Koch, M.S. Lubell, T.B. Lucatorto (American Institute of Physics, New York) p. 264

Rivarola R.D., P.D. Fainstein, V.H. Ponce (1996) in *Two-Center Effects in Ion-Atom Collisions*, ed. by T.J. Gay, A.F. Starace, AIP Conf. Proc. 362, 147

Rudd M.E., T. Jorgensen Jr. (1963) Phys. Rev. 131, 666

Rudd M.E., C.A. Sautter, C.L. Bailey (1966a) Phys. Rev. 151, 20

Rudd M.E. T. Jorgensen Jr., D.J. Volz (1966b) Phys. Rev. 151, 28

Rudd M.E., J.H. Macek (1972) Case Studies At. Phys. 3, 47

Rudd M.E. (1975) Rad. Research 64, 153

Rudd M.E., L.H. Toburen, N. Stolterfoht (1976) At. Data Nucl. Data Tables 18, 413

Rudd M.E. (1979) Phys. Rev. A 20, 787

Rudd M.E., J.S. Risley, J. Fryar, R.G. Rolfes (1980) Phys. Rev. A 21, 506

Rudd M.E., R.D. DuBois, L.H. Toburen, C.A. Ratcliffe, T.V. Goffe (1983) Phys. Rev. A **28**, 3244

Rudd M.E. (1988) Phys. Rev. A **38**, 6179

Rudd M.E., Y.-K. Kim, D.H. Madison, T.J. Gay (1992) Rev. Mod. Phys. **64**, 441

Rudd M.E., Y.K. Kim, T. Märk, J. Schou, N. Stolterfoht, L.H. Toburen (1996) *Secondary Electron Spectra from Charged Particle Interactions*, International Commission on Radiation Units and Measurements, Report 55, (Bethesda, MD)

Rudge M.R.H., M. Seaton (1964) Proc. R. Soc. (London) A**283**, 262

Rutherford E. (1911) Phil. Mag. **21**, 669

Salin A. (1969) J. Phys. B **2**, 631

Salin A. (1972) J. Phys. B **5**, 979

Salin A. (1989) J. Phys. B **22**, 3901

Sanders J., S. Datz, R.D. DuBois, S.T. Manson (1995) in *XIX Int'l Conf. on the Physics of Electronic and Atomic Collisions, Abstracts and Contributed Papers,* ed. by J.B.A. Mitchell, J.W. McConkey, C.E. Brion (Univ. Press, Whistler, Canada) p. 760

Sarkadi L., J. Bossler, R. Hippler, H.O. Lutz (1983) J. Phys. B **16**, 71

Sarkadi L., J. Bossler, R. Hippler, H.O. Lutz (1984) Phys. Rev. Lett **53**, 1551

Sarkadi L., J. Pálinkás, A.+Kövér, D. Berényi, T. Vajnai (1989) Phys. Rev. Lett. **62**, 527

Sarkadi L., M. Kuzel, L Víkor, P.A. Závodsky, R. Maier, D. Berényi, K.O. Groeneveld (1997), Nucl. Instr. Methods Phys. Res. B **124**, 335

Sataka M., K. Okuno, J. Urakawa, N. Oda (1979a) in *XIth Int'l Conf. on the Physics of Electronic and Atomic Collisions, Abstracts* (Society of Atomic Collision Research, Kyoto, Japan) p. 620.

Sataka M., J. Urakawa, N. Oda (1979b) J. Phys. B **12**, L729

Sataka M., M. Imai, Y. Yamazaki, K. Komaki, K. Kawatsura, Y. Kanai, H. Tawara (1993) Nucl. Instrum. Methods B **79**, 81

Sataka M., M. Imai, Y. Yamazaki, K. Komaki, K. Kawatsura, Y. Kanai, H. Tawara, D.R. Schultz, C.O. Reinhold (1994) J. Phys. B **27**, L171

Schiwietz G., U. Stettner, T.J.M. Zouros, N. Stolterfoht (1987) Phys. Rev. A **35**, 598

Schiwietz G. (1990) Phys. Rev. A **42**, 296

Schiwietz G., G. Xiao, P.L. Grande, B. Skogvall, R. Köhrbrück, B. Sulik, K. Sommer, A. Schmoldt, U. Stettner, A. Salin (1994) Euro. Phys. Lett. **27**, 341

Schiwietz G. (1994) (unpublished)

Schlachter A.S., K.H. Berkner, W.G. Graham, R.V. Pyle, P.J. Schneider, K.R. Stalder, J.W. Stearns, J.A. Tanis, R.E. Olson (1981) Phys. Rev. A **23**, 2331

Schneider D., M. Prost, N. Stolterfoht, G. Nolte, R.D. DuBois (1983)Phys. Rev. A **28**, 649

Schneider D., D. DeWitt, A.S. Schlachter, R.E. Olson, W.G. Graham, J.R. Mowat, R.D. DuBois, D.H. Loyd, V. Montemayor, G. Schiwietz (1989) Phys. Rev. A **40**, 2971

Schneider D., D. DeWitt, R.W. Bauer, J.R. Mowat, W.G. Graham, A.S. Schlachter, B. Skogvall, P. Fainstein, R.D. Rivarola (1992) Phys. Rev. A **46**, 1296

Schultz D.R., C.O. Reinhold (1990) J. Phys. B **23**, L9

Schultz D.R., R.E. Olson (1991) J. Phys. B **24**, 3409

Schultz D.R., R.E. Olson, C.O. Reinhold (1991) J. Phys. B **24**, 521

Schultz D.R., L. Meng, R.E. Olson (1992) J. Phys. B **25**, 4601

Schultz D.R., C.O. Reinhold (1994) Phys. Rev. A **50**, 2390

Schultz D.R., C.O. Reinhold, R.E. Olson (1996) in *Two-Center Effects in Ion-Atom Collisions,* ed. by T.J. Gay, A.F. Starace, AIP Conf. Proc. **362**, 84.

Shah M.B., H.B. Gilbody (1983) J. Phys. B **16**, 4395

Shah M.B., H.B. Gilbody (1991) J. Phys. B **24**, 977

Shakeshaft R. (1976) Phys. Rev. A **14**, 1626

Shakeshaft R. (1978) Phys. Rev. A **18**, 1930

Shanker R., U. Werner, R. Bilau-Faust, R. Hippler, U. Wille (1989) Phys. Rev. A **40**, 2335

Shingal R., Z. Chen, K.R. Karim, C.-D. Lin, C.P. Bhalla (1990) J. Phys. B **23**, L637

Shinpaugh J.L., W. Wolff, H.E. Wolf, U. Ramm, O. Jagutzki, H. Schmidt-Böcking, J. Wang, R.E. Olson (1993) J. Phys. B **26**, 2869

Skogvall B., G. Schiwietz (1990) Phys. Rev. Lett. **26**, 3265

Skutlartz A., S. Hagmann, H. Schmidt-Böcking (1988) J. Phys. B **21**, 3609

Stodden C.D., H.J. Monkhorst, K. Szalewicz, T.G. Winter (1990) Phys. Rev. A **41**, 1281

Stolterfoht N. (1971) Z. Physik **248**, 81, 92

Stolterfoht N., D. Schneider, D. Burch, H. Wiemann, J.S. Risley (1974) Phys. Rev. Lett. **33**, 59

Stolterfoht N. (1978) in *Structure and Collisions of Ions and Atoms,* Topics Curr. Phys. Vol. 5, ed. by I.A. Sellin (Springer, Berlin, Heidelberg) p. 155

Stolterfoht N., D. Schneider (1979) IEEE Trans. NS-26(1), 1130

Stolterfoht N. (1987a) Phys. Reports **146**, 315

Stolterfoht N. (1987b) *Progress in Atomic Spectroscopy,* Part D, ed. by H.J. Beyer, H. Kleinpoppen (Plenum, New York) p. 415

Stolterfoht N., D. Schneider, J. Tanis, H. Altevogt, A. Salin, P.D. Fainstein, R. Rivarola, J.P. Grandin, J.N. Scheurer, S. Andriamonje, D. Bertault, J.F. Chemin (1987) Euro. Phys. Lett. **4**, 899

Stolterfoht N. (1991) Nucl. Instrum. Methods B **53**, 477

Stolterfoht N., X. Husson, D. Lecler, R. Köhrbrück, B. Skogvall, S. Andriamonje, J.P. Grandin (1991) *XVIIth Int'l Conf. on the Physics of Electronic and Atomic Collisions, Abstract of Contributed Papers,* ed. by I.E. McCarthy, W.R. MacGillivray, M.C. Standage (Griffith Univ. Press, Brisbane) p. 393.

Stolterfoht N., D. Schneider, D. Burch, H. Wieman, J.S. Risley (1994) Phys. Rev. A **49**, 5112

Stolterfoht N., H. Platten, G. Schiwietz, D. Schneider, L. Gulyás, P.D. Fainstein, A. Salin (1995) Phys. Rev. A **52**, 3796

Stolterfoht N. (1996) in *Two-Center Effects in Ion-Atom Collisions,* ed. by T.J. Gay, A.F. Starace, AIP Conf. Proc. **362**, 163

Suárez S., C. Garibotti, W. Meckbach, G. Bernardi (1993a) Phys. Rev. Lett. **70**, 418

Suárez S., C. Garibotti, G. Bernardi, P. Focke, W. Meckbach (1993b) Phys. Rev. A **48**, 4339

Sulik B., S. Ricz, I. Kádár, G. Xiao, G. Schiwietz, K. Sommer, P. Grande, R. Köhrbrück, M. Grether, N. Stolterfoht (1995) Phys. Rev. A **52**, 387

Swafford G.L., J.F. Reading, A.L. Ford, E. Fitchard (1977) Phys. Rev. A **16**, 1329

Swenson J.K., C.C. Havener, N. Stolterfoht, K. Sommer, F.W. Meyer (1989) Phys. Rev. Lett. **63**, 35

Tanis J.A., G. Schiwietz, D. Schneider, N. Stolterfoht, W.G. Graham, H. Altevogt, R. Kowallik, A. Mattis, B. Skogvall, T. Schneider, E. Smola (1989) Phys. Rev. A **39**, 1571

Taulbjerg K. (1990a) Phys. Scr. **42**, 205

Taulbjerg K. (1990b) J. Phys. B **23**, L761

Taulbjerg K., R. Barrachina, J.H. Macek (1990) Phys. Rev. A **41**, 207

Taulbjerg K. (1991) J. Phys. B **24**, L617

Temkin A. (1959) Phys. Rev. **116**, 358

Thomson J.J. (1912) Phil. Mag. **23**, 449

Thomson G.M. (1977) Phys. Rev. **3**, 865

Thomas L.H. (1927 Proc. R. Soc. (London) **114**, 501

Toburen L.H. (1971) Phys. Rev. A **3**, 216

Toburen L.H., W.E. Wilson (1975) Rev. Sci. Instrum. **46**, 851.

Toburen L.H., W.E. Wilson (1977) *Xth Int'l Conf. on the Physics of Electronic and Atomic Collisions, Abstracts* (Commissariat A L'Energie Atomique, Paris) p. 1006

Toburen L.H. (1979a) XIth Int'l Conf. on the Physics of Electronic and Atomic Collisions, Abstracts (The Society for Atomic Collision Research, Kyoto, Japan) p. 630

Toburen L.H. (1979b) Proc. 5th Conf. on the Use of Small Accelerators, IEEE Trans. NS-26, 1056

Toburen L.H., W.E. Wilson (1979) Phys. Rev. A 19, 2214

Toburen L.H., W.E. Wilson, R.J. Popowich (1980) Radiat. Res. 82, 27

Toburen L.H., N. Stolterfoht, P. Ziem, D. Schneider (1981) Phys. Rev. A 24, 1741

Toburen L.H., R.D. DuBois, C.O. Reinhold, D.R. Schultz, R.E. Olson (1990) Phys. Rev. A 42, 5338

Toburen L.H. (1991) *Physical and Chemical Mechanisms in Molecular Radiation Biology*, ed. by W.A. Glass, M.N. Varma (Plenum, New York) p. 84

Toburen L.H., R.D. DuBois (1994) Rad. Prot. Dos. 52, 129

Tokoro N., S. Takenouchi, J. Urakawa, N. Oda (1982) J. Phys. B 15, 3737

Tokoro N., N. Oda (1985) J. Phys. B 18, 1771

Toshima N., J. Eichler (1992) Phys. Rev. A 46, 2564

Toshima N. (1994) Phys. Rev. A 50, 3940

Townsend J.S., V.A. Bailey (1922) Phil Mag. 43, 593

Trabold H., G.M. Sigaud, D.H. Jakubaßa-Amundsen, M. Kuzel, O. Heil, K.-O. Groeneveld (1992) Phys. Rev. A 46, 1270

Trautmann D., F. Rösel, G. Baur (1985) J. Phys. B 18, 1167

Urakawa J., N. Tokoro, N. Oda (1981) J. Phys. B 14, L431

van Haering H. (1985) *Charged-Particle-Interactions. Theory and Formulas* (Coulomb, Leyden)

Varga D., I. Kádár, S. Ricz, J. Végh, Á. Kövér, B. Sulik, D. Berényi (1992) Nucl. Instrum. Methods A 313, 163

Vriens L., T.F.M. Bonsen (1968) Proc. R. Phys. Soc. (London) 1, 1123

Vriens L. (1969) in *Case Studies in Atomic Collisions Physics I*, ed. by E. W. McDaniel, M.R.C. McDowell (North-Holland, Amsterdam) 335

Vriens L. (1970) Physica 47, 267

Walters H.R.J. (1975) J. Phys. B 8, L54

Wang J., J. Burgdörfer (1989) Nucl. Instrum. Methods Phys. Res. B 40/41, 65

Wang J., J. Burgdörfer, A. Bárány (1991) Phys. Rev. 43, 4036

Wang J., C.O. Reinhold, J. Burgdörfer (1991) Phys. Rev. A 44, 7243

Wang J., C.O. Reinhold, J. Burgdörfer (1992) Phys. Rev. A **45**, 4507

Wang J., R.E. Olson, H. Wolf, J. Shinpaugh, W. Wolff, H. Schmidt-Böcking (1993) J. Phys. B **26**, L457

Weiter Th., R. Schuch (1982) Z. Physik A **305**, 91

Wilson W.E., L.H. Toburen (1973) Phys. Rev. A **7**, 1535

Winter T.G. (1982) Phys. Rev. A **25**, 697

Winter T.G., C.D. Lin (1984) Phys. Rev. A **29**, 3071

Winter T.G. (1986) Phys. Rev. A **33**, 3842

Winter T.G. (1987) Phys. Rev. A **35**, 3799

Winter T.G. (1991) Phys. Rev. A **43**, 4727

Winter T.G., S.G. Alston (1992) Phys. Rev. A **45**, 1562

Winter T.G. (1993) Phys. Rev. A **47**, 264

Woerlee P.H., Yu.S. Gordeev, H. de Waard, F.W. Saris (1981) J. Phys. B **14**, 527

Wolf H., W. Wolff, J.L. Shinpaugh, H. Schmidt-Böcking (1993) Nucl. Instrum. Methods B **79**, 64

Wolff W., J.L. Shinpaugh, H.E. Wolf, R.E. Olson, J. Wang, S. Lencinas, D. Piscevic, R. Herrmann, H. Schmidt-Böcking (1992) J. Phys. B **25**, 3683

Wolff W., J.L. Shinpaugh, H.E. Wolf, R.E. Olson, U. Bechthold, H. Schmidt-Böcking (1993) J. Phys. B **26**, L65

Wolff W., J. Wang, H.E. Wolf, J.L. Shinpaugh, R.E. Olson, D.H. Jakubaßa-Amundsen, S. Lencinas, U. Berthold, R. Herrmann, H. Schmidt-Böcking (1995) J. Phys. B **28**, 1265

Wu W., R. Ali, C.L. Cocke, V. Frohne, J.P. Giese, B. Walch, K.L. Wong, R. Dörner, V. Mergel, H. Schmidt-Böcking, W.E. Meyerhof (1994) Phys. Rev. Lett. **72**, 3170

Závodszky P.A., L. Gulyás, L. Sarkadi, T. Vajnai, Gy. Szabo, S. Ricz, J. Pálinkás, D. Berényi (1994) Nucl. Instrum. Methods Phys. Res. B **86**, 175

Zouros T.J.M., D.H. Lee, P. Richard (1989) Phys. Rev. Lett. 62, 2261

Zouros T.J.M., P. Richard, K.L. Wong, H.I. Hidmi, J.M. Sanders, C. Liao, S. Grabbe, C.P. Bhalla (1994) Phys. Rev. **49**, R3155

Zouros T.J.M., C. Liao, S. Hagmann, G. Toth, E.C. Montenegro, P. Richard, E.P. Benis (1995) Nucl. Instrum. Methods Phys. Res. B **98**, 371

Index

Springer
and the
environment

At Springer we firmly believe that an international science publisher has a special obligation to the environment, and our corporate policies consistently reflect this conviction.
We also expect our business partners – paper mills, printers, packaging manufacturers, etc. – to commit themselves to using materials and production processes that do not harm the environment. The paper in this book is made from low- or no-chlorine pulp and is acid free, in conformance with international standards for paper permanency.

Springer

Printing: Saladruck, Berlin
Binding: Buchbinderei Lüderitz & Bauer, Berlin